Computational Fluid Dynamics in Drinking-Water Treatment

KWR Watercycle Research Institute Series

Computational Fluid Dynamics in Drinking-Water Treatment

Edited by

Bas Wols

Delft University of Technology

Publishing

London • New York

Published by **IWA Publishing**
Alliance House
12 Caxton Street
London SW1H 0QS, UK
Telephone: +44 (0)20 7654 5500
Fax: +44 (0)20 7654 5555
Email: publications@iwap.co.uk
Web: www.iwapublishing.com

First published 2012

British Library Cataloguing in Publication Data
A CIP catalogue record for this book is available from the British Library

Library of Congress Cataloging-in-Publication Data
A catalog record for this book is available from the Library of Congress

ISBN13: 9781780400310
ISBN: 1780400314

Contents

Acknowledgements .... **ix**

Abstract .... **xi**

Chapter 1
Introduction **1**
1.1 Hydraulics in drinking water engineering 1
1.2 Disinfection and oxidation treatment 3
1.2.1 Ozone systems 3
1.2.2 UV systems 4
1.3 Hydraulic processes 5
1.4 CFD modelling 6
1.4.1 The need for experiments 7
1.5 Aim and outline of this thesis 7

Chapter 2
Modelling tools for flow, mixing and reaction **11**
2.1 Introduction 11
2.2 Flow model 12
2.3 Tracer transport 14
2.4 Physical or chemical processes 14
2.4.1 Ozone decay model 14
2.4.2 UV irradiance model 15
2.4.3 CT value or UV dose 16
2.5 Effects on water quality 17
2.5.1 Disinfection model 17
2.5.2 Advanced oxidation model 19

2.6 Alternative disinfection models for ozone systems . . . 21
2.6.1 Maximum disinfection capacity . . . 22
2.6.2 CSTR method . . . 22
2.6.3 CT_{10} method . . . 23
2.6.4 Segregated flow analysis (SFA) method . . . 23
2.6.5 Micro-mixing analysis (MMA) method . . . 24
2.7 UV parameter study . . . 24
2.7.1 Scale factor mean dose . . . 24
2.7.2 Optimal distance from the lamp to the outer wall . . . 26
2.7.3 Approximation of dose distribution . . . 26

Chapter 3
Particle tracking . . . **29**
3.1 Introduction . . . 29
3.2 Stochastic differential equations . . . 30
3.2.1 Brownian motion . . . 30
3.2.2 Itô vs Stratonovitch . . . 30
3.2.3 Fokker-Planck equation . . . 31
3.2.4 Fickian diffusion process . . . 31
3.2.5 Advection-diffusion process . . . 32
3.3 Numerical implementation advection . . . 33
3.3.1 Euler scheme . . . 33
3.3.2 Runge-Kutta scheme . . . 33
3.3.3 Semi-analytical scheme . . . 34
3.3.4 Time step requirements . . . 35
3.3.5 Test case: rotating flow . . . 36
3.4 Numerical implementation diffusion . . . 36
3.4.1 Test case: wall treatment . . . 37
3.4.2 Test case: diffusion properties . . . 39
3.5 Test case: channel flow . . . 39
3.6 Number of particles . . . 42

Chapter 4
Ozone systems . . . **43**
4.1 Introduction . . . 43
4.2 Experiments of Leiduin ozone contactor . . . 45
4.2.1 Setup . . . 45
4.2.2 Residence time distributions . . . 46
4.3 CFD modelling of various ozone contactors . . . 46
4.3.1 Modelling conditions . . . 46
4.3.2 RTD: Eulerian versus lagrangian . . . 46
4.3.3 Validation of residence time distributions . . . 47
4.3.4 Mesh independency . . . 48
4.4 CFD results of hydraulic optimisations . . . 48
4.5 Assessment of disinfection models . . . 53

4.6 Sensitivity to kinetic parameters . . . 55
4.7 Short-circuiting . . . 57
4.7.1 Quick scan . . . 58
4.8 Conclusions . . . 59

Chapter 5
Single cross-flow UV lamp systems . . . **61**
5.1 Introduction . . . 61
5.2 Experiments of various UV lamp shapes . . . 63
5.2.1 Setup . . . 63
5.2.2 Results of flow measurements . . . 65
5.2.3 Discussion . . . 70
5.3 CFD modelling of the reference cylinder . . . 71
5.3.1 Modelling conditions . . . 71
5.3.2 Validation of velocity profiles . . . 71
5.4 Comparison between LES model and k-ε model . . . 74
5.4.1 Configurations . . . 74
5.4.2 Modelling approach . . . 75
5.4.3 Velocity fields . . . 76
5.4.4 Residence time distributions . . . 79
5.4.5 Dose distributions and disinfection . . . 80
5.4.6 Oxidation . . . 81
5.4.7 Discussion . . . 83
5.5 Conclusions . . . 84

Chapter 6
UV systems . . . **87**
6.1 Introduction . . . 87
6.2 Experiments of a bench-scale UV reactor . . . 88
6.2.1 Setup . . . 88
6.2.2 Residence time distributions . . . 91
6.2.3 Dye measurements . . . 93
6.2.4 Flow fields . . . 95
6.2.5 Lagrangian actinometry . . . 98
6.2.6 Disinfection . . . 99
6.2.7 Oxidation . . . 100
6.2.8 Discussion on the experimental results . . . 102
6.3 CFD modelling of a bench-scale UV reactor . . . 102
6.3.1 Modelling conditions . . . 102
6.3.2 Mesh independency . . . 102
6.3.3 Validation . . . 104
6.4 Design of hydraulically optimised UV reactors . . . 109
6.4.1 Overview of UV reactor designs . . . 109
6.4.2 UV dose distributions . . . 112

6.4.3 Results of energy consumption ... 116
6.4.4 Design considerations ... 119
6.5 Conclusions ... 120

Chapter 7
General conclusions ... **121**
7.1 CFD modelling aspects ... 121
7.1.1 Turbulence modelling ... 123
7.1.2 Lagrangian versus Eulerian ... 123
7.1.3 Validation techniques ... 124
7.1.4 Use of CFD models in the water industry ... 125
7.2 System design considerations ... 125
7.2.1 Ozone systems ... 127
7.2.2 UV systems ... 127
7.3 Outlook for CFD modelling in disinfection/oxidation processes ... 129

References ... **131**

Appendix A Modelling approaches ... **137**

Appendix B UV irradiation model ... **139**

List of publications ... **143**

List of symbols ... **145**

Acknowledgements

This project has been sponsored by Delft University of Technology, the joined Dutch Water Supply Companies and KWR Watercycle Research Institute, and for the last two years co-sponsored in the TTIW-cooperation framework of Wetsus, centre of excellence for sustainable water technology (http://www.wetsus.nl). Wetsus is funded by the Dutch Ministry of Economic Affairs. The participants of the research theme clean water technology are thanked for the fruitful discussions and their financial support.

Abstract

Computational fluid dynamics (CFD) uses advanced numerical models to predict flow, mixing and (bio)-chemical reactions. In drinking water engineering, CFD is increasingly applied to predict the performance of treatment installations and to optimise these installations. In this thesis the hydraulics in ozone and UV systems are studied by means of CFD. A lack of understanding of the hydraulics in drinking water treatment systems has resulted in suboptimal design of installations. The formation of unwanted disinfection-by-products and the energy consumption or use of chemicals is therefore higher than necessary. The aim of this work is to better understand the hydraulic and (bio)-chemical processes in drinking water treatment installations using experimental and numerical techniques. By combining these techniques, CFD modelling is further developed as a tool to evaluate the performance of these installations. This leads to new insights in the applicability of models in ozone and UV systems, and new insights in design concepts of these systems.

As a starting point of the CFD modelling, a flow model is solved to determine the velocity fields and turbulent properties inside a system. Using the data from the flow model, individual particles are tracked, which represent micro-organisms. For the particle tracking technique, special attention is required to prevent particles from, unrealistically, accumulating at regions with low diffusion and to prevent them from virtually crossing the walls. The particle model is combined with a disinfectant or oxidant calculation (ozone concentration, UV irradiance or hydrogen peroxide). The disinfectant or oxidant can be integrated over the particle's trajectory, resulting in the dose (CT value or UV dose). By using a large number of particles, a distribution of doses is obtained, which determines the disinfection or oxidation performance. For a good design of ozone and UV systems, the dose distribution needs to be as narrow as possible. The residence time distribution (RTD) is insufficient to characterise the performance of a UV system, because it does not account for the spatial distribution of UV intensity. For the single cross-flow UV lamp systems, short residence times were compensated by high UV irradiances, so that systems with a wider RTD showed a better disinfection performance than systems with a smaller RTD.

The most challenging task is to model the flow field accurately because simplifications are needed to describe the turbulence. Different modelling approaches were considered to cope with turbulent flows. A large-eddy simulation technique (LES model) as well as a Reynolds-averaged Navier-Stokes (RANS) approach with a k-ε turbulence model (k-ε model) were evaluated. The LES model resolves the large-scale time-varying turbulent motions, whereas the RANS approach solves a time averaged velocity field and models the turbulent fluctuations by means of an artificial viscosity term. The most commonly used model in practice is the standard k-ε model, because accurate results are obtained at reasonable calculation times (of the order of one day). The modelling approaches were evaluated for a single cross-flow UV lamp system and a bench-scale UV reactor. We showed that in comparison with experimental data recirculations were better represented by the LES model, which could have a large impact on the disinfection prediction. Especially for a single cross-flow UV lamp system with baffles, large recirculations could develop, and the k-ε model largely overestimated the disinfection. For UV reactors used in practice the geometry is more confined, so that large recirculations can not develop. The differences between both model approaches are therefore expected to be smaller. This was demonstrated for the bench-scale UV reactor, where the k-ε model predicted a 10% higher disinfection level than the LES model. The choice between a RANS model and a LES model is a trade-off between accuracy and available computational resources. To evaluate full-scale UV systems within a few days, one is still obliged to use RANS models, but the limitations of this modelling approach should be kept in mind.

The hydraulics in water treatment installations can be improved by streamlining the flow - reducing recirculations - and by increasing the degree of mixing. A high mixing ensures that regions of low doses are mixed with regions of high doses, resulting in a narrower dose distribution. A common measure to improve the hydraulics is placing baffles, which increase the mixing and may streamline the flow. However, the baffles also cause a recirculation, so the position of the baffle should be chosen carefully. We showed that placing additional baffles in an ozone contactor resulted in higher disinfection levels, because the large recirculations were reduced. For a single cross-flow UV lamp system, only baffles that were placed a little downstream of the UV lamp resulted in a higher disinfection performance. In addition, a wide range of UV reactors were assessed by CFD. Although the input parameters such as lamp power, flow rate and transmittance were the same, the dose distributions differed greatly for the various UV systems. The best results were obtained by an annular system, designed as a static mixer with a lamp in the middle, which came closest to the theoretically perfect hydraulic system (within 10%). Baffles placed in different directions in this annular system increased the mixing and forced all the particles to come close to the lamp, resulting in a narrow dose distribution.

In conclusion, CFD modelling proves to be a powerful tool to understand the hydrodynamic and (bio)-chemical processes in drinking water systems. If applied properly, accounting for the complex turbulent motions and validated by experiments, this tool leads to a better design of UV reactors, ozone systems and other systems dictated by hydraulics.

Bas Wols
June 2010

Chapter 1

Introduction

In the first Chapter a general introduction about hydraulics and CFD modelling is given, which is applied to processes in drinking water treatment, such as disinfection and oxidation.

1.1 HYDRAULICS IN DRINKING WATER ENGINEERING

People regard drinking water as one of the basic elements of life and, therefore, want to have water that is safe and easily available against low costs. The transport of water is an inevitable prerequisite, since clean water has to be delivered at home. In the Netherlands nowadays, this means that the water has to be transported from the source (groundwater or surface water) towards a treatment plant, and from the treatment plant through the distribution system towards the end-users. Understanding hydrodynamics is therefore essential in order to design and operate an efficient drinking water system. Moreover, within the treatment plant, processes occur that are very sensitive to the hydraulics, such as the inactivation of pathogenic micro-organisms (disinfection). The pathogen has to be in contact with a disinfectant for a sufficient time (Smeets, 2008), which depends on the movement of the water inside such a system.

In the Netherlands the water is treated through an approach, which applies multiple barriers against pathogens. As an example of drinking water engineering the Leiduin water treatment plant in Amsterdam is considered (Figure 1.1). First, river water is pre-treated by coagulation, sedimentation and filtration, after which it is transported to the dunes. The water is extracted from the dunes and treated by aeration, rapid sand filtration, ozonation, softening, activated carbon filtration and slow sand filtration. The water is stored in a clear water reservoir, before it is transported through the distribution network. During the (pre-)treatment in these plants processes, where the hydraulics play an important role, are:

- *Sedimentation*. In a settling tank flocks settle to the bottom. The efficacy of settling depends, next to the characteristics of the flocks, on the flow velocities in the tank (van der Walt, 2002).
- *Filtration*. During the backwash, the filter is cleaned by an upwards water flow, usually combined with air flushing. This results in a complex multi-phase flow of water and air, where the hydraulics define the efficacy of the cleaning (Schwager and Boller, 1997). The water that is infiltrated in the dunes flows slowly through the pores of the sand before it is extracted. Also, for the production of drinking water, groundwater can be extracted from wells, which influences the

natural groundwater flows. Knowledge on these flows is important, both regarding quantity (draw down) as well as quality (pollution).

- *Mixing and division of flows.* Mixers are important to mix chemicals with the water, for example dosing of coagulants, oxidants, etcetera. The turbulent flows in the mixer determine the efficacy of the mixer. Flow dividers are used in the water treatment plant to divide the flow into parallel treatment lanes, and the local velocities and pressure gradients determine the amount of water that enters each treatment lane.
- *Ozonation.* Ozone contactors are widely used for disinfection and oxidation. The hydraulics are of great importance for the disinfection efficacy of ozone contactors (Rietveld, 2005; van der Helm, 2007; Smeets, 2008), which will be topic of this thesis. The hydraulics of the Leiduin ozone contactor are addressed in Chapter 4.
- *Softening in pellet reactors.* A complex multi-phase flow (solid-liquid) develops in pellet reactors, where the water flows interact with the moving pellets. Understanding these processes helps to improve the efficacy of the reactor (van Schagen, 2009; Dixon and Nijemeisland, 2001).
- *Storage in clear water reservoirs.* Hydraulics in the clear water reservoirs determine the 'age' of the water in the tank, which is important for the quality of the water flowing from the tank to the distribution network (van der Walt, 2002).
- *Resuspension in the distribution network.* The hydraulics (i.e., turbulent intensities) determine if sediments settle in the distribution network or if settled sediments are resuspended. These processes are important for discoloured water problems at the tap (Vreeburg, 2007).

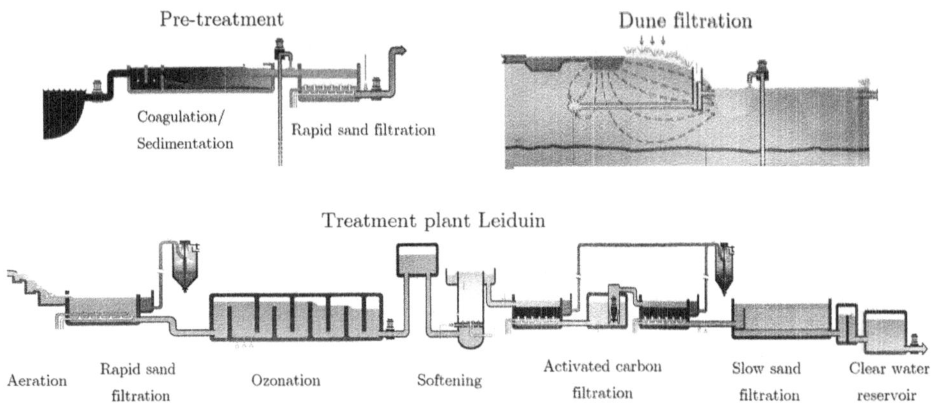

Figure 1.1 Treatment scheme of the Leiduin water treatment plant (Waternet).

Some other drinking water treatment processes, which are not used in the above mentioned example, are also largely influenced by the hydraulics:

- *Ultraviolet (UV) radiation.* UV systems are used to inactivate micro-organisms or to degrade organic substances (often in combination with hydrogen peroxide). The hydraulics are very important for the disinfection and oxidation performance, which will be topic of this thesis in Chapter 5 and 6.
- *Membrane filtration* (van der Meer, 2003; Verliefde, 2008). The (turbulent) flow in and around the membrane influences the water flux through the membrane, a higher degree of turbulence results in a better mixing and therefore a better performance of the system (Verberk, 2005). Also, stagnant

zones may increase the fouling on the membranes. Optimising the hydraulics by increasing the mixing and reducing stagnant zones may result in better designs of membrane filters (Ghidossi et al., 2006).

Understanding the hydraulic processes in the above mentioned systems is of interest for the more theoretical mind, i.e., for a better notion of the actual process, and for the more practical mind, i.e., to optimise the process and reduce costs or energy consumption. The theoretical investigations of the hydraulics and (bio)-chemical processes lead to improved modelling of these processes. Moreover, the experimental and numerical results lead to practical considerations, which are useful for the design of treatment installations. In this thesis, we will focus on the disinfection and oxidation processes, but the numerical tools described here could be expanded to the other water treatment processes as well.

1.2 DISINFECTION AND OXIDATION TREATMENT

Disinfection of drinking water is required to prevent the outbreak of water-borne diseases. The disinfection aims to inactivate the micro-organisms, so that they can no longer infect humans. Different treatment steps are available to disinfect the water, such as chlorination, ozonation or UV irradiation. In the Netherlands, ozonation and UV irradiation are both used for disinfection, whereas treatment with chlorine is no longer applied due to formation of unwanted disinfection by-products. According to the Dutch drinking water decree (VROM, 2001), the drinking water is considered safe when the infection risk does not exceed 10^{-4} pppy, which means that 1 out of 10 000 people per year may become infected from drinking potable water. These strict requirements often ask for an inactivation target, measured in decimal elimination (*DE*), of log 3 (99.9%) or log 4 (99.99%) (Sommer et al., 2008), or sometimes even more (Smeets, 2008), depending on the quality of the source and disinfection by other drinking water treatment steps. To reach the disinfection target, each micro-organism needs to be exposed to a sufficient high amount of disinfectant for a sufficient length of time. The disinfectant dose (also called CT value for ozone processes, and UV dose for UV processes), defined as the contact time of a micro-organism with the disinfectant, is mainly determined by the hydraulics. Due to differences in water velocities the micro-organisms follow different trajectories in the installation. Therefore, in combination with the spatial distribution of disinfectant, each micro-organism receives a different disinfectant dose. The distribution of disinfectant doses for individual micro-organisms largely determines the efficacy of the disinfection step.

Oxidation is used to degrade chemical compounds. These compounds are mainly organic pollutants, such as pesticides, endocrine-disrupting compounds, algae toxins and pharmaceuticals (Kruithof et al., 2007). Ozone treatment is used as an oxidation process (besides disinfection), and UV in combination with hydrogen peroxide is used as an advanced oxidation process. The removal target is less severe for oxidation than for disinfection, a criterion of around 80% degradation is not uncommon (Kruithof et al., 2007).

1.2.1 Ozone systems

Ozone systems are widely used for disinfection purposes, but also to oxidise organic micro-pollutants and to stimulate biological activity in the activated carbon filters (Rietveld et al., 2008). Other additional benefits of ozone treatment are: oxidation of reduced metals, reduction in taste, odour and colour of the water. An ozone system consists of a dosing chamber and some reaction chambers. The ozone gas is usually dosed to the water by means of a bubble column, where the ozone is dissolved in the water. The bubble column is followed by a number of contact chambers to ensure sufficient reaction time. The best disinfection results are reached when the dissolved ozone is perfectly mixed and the flow in the contact chambers resembles a plug flow. These criteria are often not met, which could result in reduced performance of the ozone contactor (Smeets et al., 2006). In the past, little attention was paid to the flow in ozone systems. It was often treated as a black box, neglecting the local flow phenomena. But the complex flow patterns with its effect on the disinfection capacity need to be taken into account. Due to the suboptimal hydraulics, some parts of the water receive less ozone than other parts of the water. Short-circuit flows in the ozone contactor are mainly responsible for volumes of water that receive low CT values.

One of the major issues for ozone systems is the formation of bromate, which is a suspected carcinogenic substance. Bromate is formed when bromide containing water is treated by ozonation (von Gunten, 2003a). The bromate formation depends on the bromide concentration in natural water and the CT value. At the Leiduin water treatment plant (Waternet) pathogen infection risk analysis indicated that improved disinfection is required (Kappelhof et al., 2007; Smeets, 2008). The formation of disinfection by-products prohibits an increase in ozone dosage, so an optimisation of the hydraulics is required to improve disinfection capacity. Understanding the hydraulics of ozone systems is therefore of utmost importance.

1.2.2 UV systems

Disinfection by UV radiation and advanced oxidation by UV radiation in combination with hydrogen peroxide are increasingly being applied in water treatment. A breakthrough - using UV as a disinfection technology - resulted from the work of Bolton et al. (1998) and was confirmed by Craik et al. (2000, 2001) who proved that low pressure UV radiation results in a very effective inactivation of *Cryptosporidium parvum* and *Giardia lamblia* at low UV doses. UV disinfection has also been included as an accepted process in "the long term 2 enhanced surface water treatment rule" (USEPA, 2006b), which describes the regulations for the drinking water treatment in the US. The inactivation of micro-organisms by UV radiation is caused by damaging of the DNA of the cell (or virus), which prevents the micro-organism from multiplying. The maximum absorbance of DNA occurs at a wave length of around 260 nm (Hijnen et al., 2006). Advanced oxidation by UV and hydrogen peroxide is successful due to the formation of hydroxyl radicals that degrade organic contaminants (de Laat et al., 1999).

Two techniques of UV irradiation are used in water treatment: a monochromatic or low pressure UV lamp and a polychromatic or medium pressure UV lamp (Sommer et al., 2008). The low pressure lamp emits radiation at a very narrow range of wavelengths of around 254 nm (Figure 1.2). The DNA of micro-organisms is most effectively damaged at these wavelengths, so that the low pressure lamps are often used for disinfection purposes. The medium pressure emits radiation at a broad range of wavelengths, mostly between 200 nm and 300 nm. The formation of hydroxyl radicals from hydrogen peroxide by the absorption of photons is more efficient at lower wavelengths (see Figure 1.2). However, the absorption of UV radiation in the water is higher at these lower wave lengths. We will focus on the low pressure UV lamps for disinfection and oxidation, but the tools developed here could be expanded to UV lamps with a broad wave length spectrum.

Another point of concern is the power consumption of UV lamps, since only approximately 30% of the electrical energy is transformed into UV radiation. Especially for advanced oxidation, high amounts of electrical power are needed to degrade organic substances in the water. The effectiveness of the disinfection and oxidation depends on the spatial distribution of UV radiation and on the hydraulics in the system. Hydraulic processes determine the pathways and residence times of micro-organisms as well as the production of hydroxyl radicals. A large variation exists in the geometries of UV reactors, which may result in completely different hydraulic processes occurring within each system, and, therefore, disinfection and oxidation performance can be different.

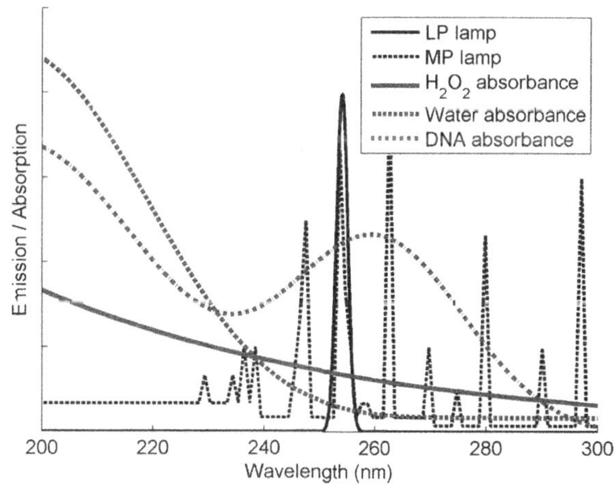

Figure 1.2 Emission spectrum of a low pressure (LP) and medium pressure (MP) UV lamp; absorption of UV by hydrogen peroxide, by water, and by DNA of micro-organisms.

1.3 HYDRAULIC PROCESSES

The term hydraulic or hydrodynamic process – a rather general term – is used here to represent all the processes that are related to the movement of fluids. One of the most complex issues is turbulence, a complete description of turbulent flow remains one of the unsolved problems in physics. In contrast with laminar flows, which are smooth, orderly and flowing in parallel layers, turbulence is characterised by irregular, chaotic, rotational, and dissipative behaviour; eddies are formed at many different length scales (Davidson, 2004). In water treatment installations, the flows are usually turbulent due to the large dimensions (of the order of meters) and high water velocities (ranging from 0.1 m/s to 2 m/s). Because of the formation of eddies, turbulent flows have a high degree of mixing, which is useful in practice. For example, static mixers increase the turbulent intensities to create a high level of mixing. The mixing is usually described as an eddy viscosity process, which is in fact not a proper physical concept because it is a property of the flow and not of the fluid, such as molecular viscosity. Nevertheless, describing the mixing of turbulent motions by an eddy viscosity coefficient is very successful and is used in many applications (Launder and Spalding, 1974). Some computational fluid dynamics (CFD) models have turbulence models (e.g., k-ε model) that use this approach. An example of a turbulent flow is shown in Figure 1.3, where the turbulent wake behind a cylinder is visualised by dye (potassium permanganate). The different length scales of the turbulent eddies can be clearly seen.

Another important hydraulic property is a recirculation zone or separation bubble, also called wake region. A recirculation zone develops when the flow becomes detached from a surface, which occurs for example downstream of a baffle or cylinder. As a result, a large separation zone may develop with reversing flow directions. Residence times are large for the water trapped inside such a separation bubble. The size of the recirculation zone is sensitive to the upstream flow conditions and Reynolds number (Hanjalić and Jakirlić, 1998).

Short-circuiting occurs when parts of the water travel rapidly through the reactor, and, as a result, these parts receive insufficient disinfectant dose. These short-circuits are unfavourable for the disinfection performance because a small volume with insufficient disinfection may impact the total disinfection of the system considerably. For example, a short-circuit flow of 1% of the total flow rate limits the inactivation to a maximum of log 2.

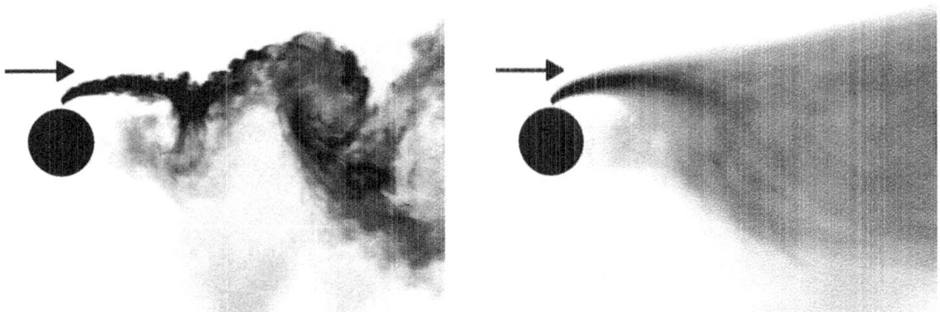

Figure 1.3 Turbulent wake formation for the flow around a cylinder, the flow (coming from the left) is visualised by coloured dye. An instantaneous dye concentration (left panel) and a dye concentration averaged during 2 s (right panel) are displayed.

The residence time distribution is a well-known characterisation of the hydraulics. This distribution shows the differences in residence times between two points, normally the inlet and outlet of a flow-through system. Each residence time represent the travelling time of a water element in the installation, which indicates the presence of short-circuits and recirculation zones.

1.4 CFD MODELLING

Ongoing development of computers and software technology allows detailed modelling of the water flow through reactors and the influence of the hydraulics on physical, chemical and microbiological processes taking place in the reactors. Computational fluid dynamics (CFD) is increasingly applied to investigate and optimise hydraulic processes in a wide range of applications. For example, prediction of meteorological systems (Kovalets et al., 2008), wind flow and dispersion in urban environments (Solazzo et al., 2009), flooding by rivers (Elíasson et al., 2007) or predicting the efficacy of drinking water installations (Craig et al., 2002; Liu and Ducoste, 2006). Using CFD models enables us to predict the flow and (turbulent) mixing processes at different levels of detail.

Several studies show that CFD serves as a powerful tool to optimise ozone contactors (Cockx et al., 1999; Huang et al., 2004; Li et al., 2006; Zhang et al., 2007). They showed that the hydraulics and/or microbial inactivation are improved by changing the contactor geometry. The disinfection performance of UV reactors were also evaluated using CFD (Lyn et al., 1999; Wright and Hargreaves, 2001; Ducoste et al., 2005; Sozzi and Taghipour, 2006; Pan and Orava, 2007). For disinfection, the spatial distribution of disinfectant (dissolved ozone or UV irradiation) was calculated. Combined with particle tracking techniques valuable statistics, such as residence time distribution and disinfectant contact time distribution, were obtained. The above mentioned authors found differences in the disinfection performance between the various reactors, which were related to the hydraulics of the systems. Modifications to the geometry of the reactor, such as baffle plates or rings, can result in an increase in disinfection capacity (Blatchley III et al., 1998; Chiu et al., 1999). For oxidation processes, the work of Sozzi and Taghipour (2007) as well as Alpert et al. (2007) are mentioned. These researchers predicted the conversion of chemical compounds in annular reactors by UV in combination with hydrogen peroxide. Sozzi and Taghipour (2007) found that a reactor with hydraulics closer to a plug flow reactor resulted in a higher conversion of chemical compounds.

The most challenging part of the simulation is to predict the flow fields correctly. The complex turbulent flows require a detailed numerical modelling to resolve all the relevant turbulent scales, which results in long and expensive computations. These kinds of simulations, known as direct numerical simulations, are impossible to perform with the present day computers - even for a bench-scale installation the simulation time would typically be of the order of 10-100 years (Davidson, 2004). A more simplified but still accurate simulation is the large-eddy simulation (LES) technique, which resolves only the larger turbulent eddies (Moeng, 1984). The LES still requires extensive computational resources and is therefore not often used in the engineering practice. Simplified turbulence models (such as the k-ε model), using the Reynolds-averaged Navier-Stokes (RANS) equations, are more often used in practice. These types of models use an averaged velocity field and the concept of eddy viscosity, which models the fluctuating turbulent motions as an averaged viscosity coefficient. Consider again Figure 1.3, which shows that a time-averaged picture of the dye concentration (right panel) is a poor measure of the instantaneous dye concentration (left panel). As a consequence, the flow fields are often predicted less accurately, especially around complex geometries. A system with UV lamps placed perpendicular to the flow direction is such a complex geometry, where the lamp forms an obstacle in the flow producing a complex turbulent flow pattern. Several studies show that the computed velocity fields in UV reactors may deviate from the measured velocity field (Lyn et al., 1999; Liu et al., 2007). In this research, more advanced CFD models, such as the large-eddy simulation, are also explored to assess the accuracy of the available modelling techniques with respect to the flow fields, and to show how the inaccuracies in the calculated flow field influence the disinfection prediction (Chapter 5 and 6).

1.4.1 The need for experiments

A comparison between experimental and numerical work is shown in Table 1.1. Experiments determine the flow and (bio)-chemical properties of a system, whereas CFD models predict them. Experiments are mostly conducted at a laboratory scale. Although all the required physical and (bio)-chemical processes are present in the experiment, due to practical limitations, the measuring devices can only determine a small number of physical or (bio)-chemical variables. In the CFD model, a full-scale installation can be modelled more easily and a large number of physical and (bio)-chemical variables can be predicted at a high resolution, although not all the relevant length and time scales can be resolved in the CFD model. Also, geometric modifications can be assessed easily, which makes CFD modelling a powerful tool for optimisation. In contrast, modifying the geometry in an experiment is time-consuming and expensive. The range of operating conditions in the experiment is limited, whereas the CFD model is more flexible in applying different operating conditions, including the real (full-scale) conditions that are often not possible in a down-scaled experiment. Conducting the experiments is usually expensive and time-consuming, whereas simulating with CFD is relatively cheap (Hilgenstock and Ernst, 1996), although it may still be time-consuming. In the end, despite the advantages of the CFD model, the modelling results are less accurate than the experimental results, since assumptions are made to model the turbulence, and only a limited number of (bio)-chemical reactions can be resolved. Therefore, experiments are still needed for a better understanding of the physical and chemical processes as well as for validation of the CFD models. The combination of CFD with experiments is therefore recommended, because it provides a complete picture of the hydraulic processes in an installation. Moreover, once the model is validated for a certain installation, the CFD model can be used to optimise the installations with respect to energy consumption, use of chemicals, and disinfection or oxidation efficacy.

1.5 AIM AND OUTLINE OF THIS THESIS

The aim of this work is to better understand the hydraulic and (bio)-chemical processes in drinking water treatment installations using experimental and numerical techniques. By combining these techniques, computational fluid dynamics modelling is further developed as a tool to evaluate the performance of these installations. This leads to new insights in the applicability of models in ozone and UV systems, and new insights in design concepts of these systems. The major focus is on the hydraulics, studied by various experimental techniques and CFD modelling. It is also investigated in which detail the velocity fields need to be resolved by the numerical model in order to obtain an accurate prediction for the disinfection and oxidation. The validation of the CFD models for different physical and (bio)-chemical processes is therefore a crucial part.

Table 1.1 Comparison between experimental and numerical techniques.

Experiments	CFD models
All the relevant length and time scales	Only the larger length and time scales
All the relevant (bio)-chemical reactions	Small number of (bio)-chemical reactions
Limited number of variables can be measured simultaneously	Larger number of variables can be computed simultaneously
Laboratory-scale	Full-scale
Fixed geometry	Geometric modifications
Limited range of operating conditions	All realistic operating conditions
Expensive and time-consuming	Relatively cheap
Very accurate	Less accurate

The techniques used to numerically model the flow, mixing and reaction are described in **Chapter 2**. These modelling tools form the starting point of the CFD predictions, and can in general be used for systems where hydraulics processes combined with (bio)-chemical reactions play an important role. A particle tracking technique is an essential part of the CFD modelling, and is treated more extensively in **Chapter 3**. This technique models the movement of passive particles, i.e, particles that move entirely with the flow. Different numerical schemes are discussed to model the particle movements precisely and to prevent particles from crossing closed boundaries. Using these modelling tools, the performance of ozone and UV systems with respect to disinfection and oxidation are predicted in the following chapters. These chapters have a similar structure, starting with the presentation of measurements that are used for a better understanding of the hydraulics and for validation of the numerical models. The results of the measurements are followed by the results of the CFD model, mesh independency tests and validation tests. Using the validated model, the design of new or optimised systems are investigated by CFD. Also, different modelling techniques for the various applications are assessed.

The first application of the CFD model is the simulation of an ozone contactor, described in **Chapter 4**. The residence time distributions predicted by the CFD model are validated by measurements in a full-scale ozone contactor (Leiduin water treatment plant, Waternet). Furthermore, the hydraulics are optimised using the CFD model by adjusting the geometry. Because various techniques exist that predict the disinfection of ozone systems, ranging from analytical models to CFD models, a thorough assessment is made of all the disinfection prediction methods using the CFD results of the different ozone contactor geometries.

Large-scale UV systems usually contain a number of cylindrical UV lamps placed perpendicular to the flow direction. In these systems the lamps cause a distortion of the otherwise straight flow. An important feature is the recirculation zone that develops downstream of the circular UV lamp, where the residence times are larger than in the main flow region. The difficulties regarding the modelling of the wake region are an important challenge for the CFD models. Therefore, the wake region of a single UV lamp is thoroughly investigated in **Chapter 5**, experimentally and numerically. A single cross-flow UV lamp is considered, and hydraulic modifications aimed at reducing the recirculation zone are investigated experimentally. For the numerical modelling, both the standard k-ε model and the LES model are evaluated. The experimental results provide an extensive validation for the CFD modelling. By considering a small part of a full-scale UV reactor in detail, such as the flow field around a single UV lamp, the reliability of different CFD models is further clarified. The aim is to assess the accuracy of the available modelling techniques with respect to the flow fields, and to explore how the inaccuracies in the calculated flow field influence the disinfection prediction.

Next to the simplified cross-flow UV system, a bench-scale UV reactor, which is more representative for UV reactors used in practice, is investigated experimentally and numerically in **Chapter 6**. A large set of experimental data - flow fields, residence time distributions, dose distributions and microbial inactivation - allows to make a complete validation of the bench-scale reactor. Using the validated model, a systematic approach for the design of UV reactors is followed. A large set of different UV reactors is assessed. Also, design aspects such as placing mirrors, determining the lamp size and placing reactors in series are considered. The various designs are evaluated on the basis of the following criteria: disinfection performance, oxidation performance and energy consumption. The potential for energy reduction by optimising the hydraulics for existing UV reactors and new reactor designs is explored as well.

In **Chapter 7** the results are summarised and conclusions are drawn.

Chapter 2

Modelling tools for flow, mixing and reaction

This chapter provides an overview of the tools to model flow, mixing and reaction, which are required to predict the performance of water treatment systems.

2.1 INTRODUCTION

Computational fluid dynamics are used to model the disinfection and oxidation performance of ozone or UV systems. A number of sub models needs to be solved to come to this prediction (Figure 2.1). There is a hierarchy in the modelling, each sub model requires information from the model(s) above.

A flow model is the starting point of the CFD calculation, which determines the velocity fields and turbulent properties inside a system. The most challenging task is to model the flow field accurately because simplifications are needed to describe the turbulence. Different modelling approaches are described in §2.2 to cope with turbulent flows. Experimental validation is then essential to get to a proper numerical modelling, which can for example be done by laser Doppler anemometry (LDA), particle image velocimetry or acoustic Doppler velocimetry. In this thesis, the hydraulics of different UV systems were validated by LDA measurements.

From the results of the flow model, a tracer transport model is used to determine a residence time distribution (§2.3). In the Eulerian approach concentrations are calculated on a fixed computational mesh, whereas in the Lagrangian approach individual particles are tracked. The latter approach will be further discussed in Chapter 3. Experimental validation of the calculated residence times is conducted by tracer experiments, where a passive tracer (for example sodium chloride, fluoride or dye) is released into the system. These kinds of measurements were conducted here for an ozone system as well as for a UV system.

The next step is to describe the physical and chemical processes that are responsible for the disinfection or oxidation, explained in §2.4. The processes used here are UV irradiation and ozonation. In combination with particle tracking techniques, the physical or chemical variable is integrated over the particle path, which results in a dose (UV dose or CT value) for each particle. An example of a validation technique for the predicted doses is Lagrangian actinometry, where the UV doses received by microspheres are measured (Blatchley III et al., 2008).

Next, the effects on the water quality are evaluated (§2.5) by calculating the inactivation of micro-organisms or conversion of a chemical compound. The predicted inactivation or oxidation can be validated by measurements of the inactivation of micro-organisms (biodosimetry) or degradation of compounds. These measurements were conducted for a bench-scale UV reactor, described in Chapter 6.

Models	Variables	Experimental validation
FLOW Navier-Stokes solver Turbulence modelling	Velocities, Turbulent properties	Laser Doppler anemometry Particle image velocimetry Acoustic Doppler velocimetry
TRACER TRANSPORT Advection-diffusion Particle tracking	Concentrations, Residence time distribution	Tracer (salt, dye etc.)
PHYSICAL OR CHEMICAL PROCESSES Advection-diffusion-decay Irradiation model	Concentrations (ozone, peroxide), UV light irradiation, CT values, UV doses	Ozone concentration (sampling) Lagrangian actinometry
EFFECTS ON WATER QUALITY Kinetic modelling	Inactivation of micro-organisms Conversion of chemical compounds	Biodosimetry Oxidation measurements

Figure 2.1 Overview of the sub models required to predict the performance of drinking water treatment systems.

Also, instead of using a complex and expensive CFD model, more simplified models exist to predict the disinfection or oxidation. These models are discussed in §2.6 and an assessment of these models is made for the ozone contactor in Chapter 4. Finally, typical parameters used for the design and operation UV systems are introduced in §2.7.

2.2 FLOW MODEL

The starting point of the modelling is the flow model. The equations that describe the flow of fluids are the continuity equation (mass conservation) and Navier-Stokes equations (momentum conservation), which is derived from Newton's second law. The velocities, pressures and turbulent properties are solved on a computational mesh.

Modelling turbulent flows is a complicated issue because turbulent motions stretch out over a large range of length scales. Solving all the turbulent length scales on the computational mesh requires very high computational times. Turbulence models are therefore used to reduce computational times at a cost of less accurate solutions. Therefore, two modelling approaches were used that differ by the amount of resolved details of the turbulent motions (Appendix A). One type solves the Reynolds-averaged Navier-Stokes equations (RANS) and uses a specific turbulence model that captures the time-varying turbulent motions in averaged parameters. Here, a 2 parameter turbulence model is used: the standard k-ε model, which is the most well-known turbulence model. From now on, the RANS approach with the k-ε turbulence model will be referred to as k-ε model. This model is most often used for engineering applications. The time-varying turbulent motions are captured in the averaged parameters k, expressing the turbulent kinetic energy, and ε, expressing the dissipation of turbulent kinetic energy. Results can be obtained at acceptable computational costs (e.g., within 1 or 2 days on a present day workstation). For that reason, these types of models are most often used for the design or optimisation of large-scale installations, such as UV reactors. The standard k-ε model is used, which uses the commonly used values for the constants in the model obtained from experiments (Launder and Spalding, 1974).

Another approach, the large-eddy simulation technique (Moeng, 1984) has much higher computational costs (e.g., within 1 or 2 weeks on 4 computational nodes), but more detailed and accurate results are obtained, because the time-varying turbulent motions are resolved. The large-eddy simulation, from now on LES model, resolves the large-scale turbulent eddies (length scale of the order of the mesh size and larger), whereas the small-scale turbulent eddies are modelled using a subgrid-scale model. A filtered Navier-Stokes equation is therefore solved with an additional subgrid-scale stress term, which accounts for the unresolved turbulent length scales by means of an eddy viscosity. The commonly used Smagorinsky subgrid-scale model was implemented here (Appendix A), where the Smagorinsky constant was taken as $C_s = 0.1$. Complex geometries can be implemented by an immersed boundary method (IBM), which forces velocities to be zero at the walls. This wall can be of arbitrary shape and does not have to be grid-aligned. More details on the LES model and the applied IBM method are provided by van Balen et al. (2009). A LES model is often used as a research tool for a better understanding of local transport mechanisms.

The choice for a k-ε model or a LES model is a trade-off between accuracy and computational speed. The differences between a LES model and a k-ε model are further clarified in Figure 2.2 for the flow around a cylinder. The LES model calculates time-varying velocity fields, which are continuously changing due to the turbulent motions; a snap-shot is shown in the upper panel of the Figure. The k-ε model only calculates a stationary mean velocity field, which always differs significantly from the unsteady velocity field occurring in reality. Both methods were used here and the differences between the methods in the application of a UV system were explored (Chapter 5 and 6).

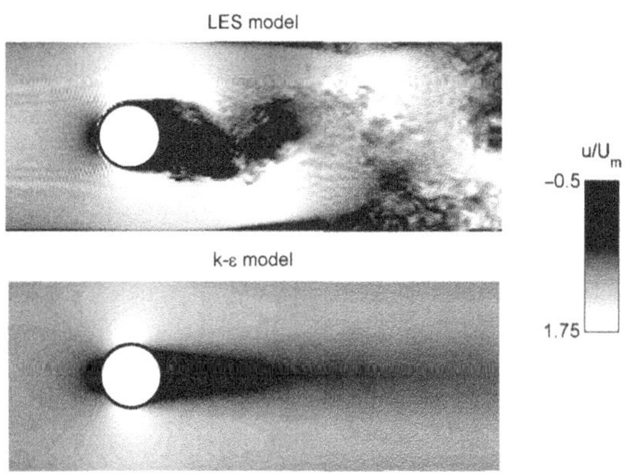

Figure 2.2 Differences in velocity field between the LES model (upper panel) and k-ε model (lower panel).

Different boundary conditions can be imposed in the flow model: at the inlet a velocity profile is imposed; at the outlet a zero pressure is imposed; solid walls are treated as logarithmic boundaries: at the wall a no slip condition is described, whereas at the first grid point a velocity in correspondence with a logarithmic velocity profile is imposed; free surfaces are treated as a moving free surface or as a rigid lid (free slip condition).

Different discretisation methods are available, for example the finite-element method, finite-volume method and finite-difference method. In principle all the methods could be used: each method has its advantages and disadvantages with no clear preference for one of the methods. We use a finite-element and a finite-difference model, for which the choice was made based upon availability of these models, either in a commercial code or in an in-house developed code. Three types of CFD codes were used here:

- A finite-element code, Finlab (Labeur and Wells, 2007), with a standard k-ε turbulence model. This model was used for 2D applications because the complex 3D geometries are difficult to handle with this model. Triangular elements were used.
- A commercial available finite-element package, COMSOL v3.5, with a standard k-ε turbulence model. This model can handle complex 3D geometries. The flow domain was discretised with tetrahedrons.
- A finite-difference code (van Balen et al., 2009) was used for the LES model. The equations were solved on a structured Cartesian grid using finite differences. A staggered grid was used, where the scalars (pressure, eddy viscosity) were solved in the middle of the volume and the vectors (velocities) were solved at the boundaries. Complex geometries can be handled using the IBM method.

2.3 TRACER TRANSPORT

In the Eulerian approach, the spreading of the tracer is modelled with an advection-diffusion equation. The advection follows from the velocities calculated by the flow model. In the case of a k-ε model, the eddy viscosity (ν_t, [m^2/s]) is derived from the turbulent kinetic energy (k, [m^2/s^2]) and turbulent dissipation (ε, [m^2/s^3]):

$$\nu_t = \frac{c_\mu k^2}{\varepsilon}, \tag{2.1}$$

where the standard value of 0.09 is used for constant c_μ (Launder and Spalding, 1974). It is assumed that the turbulent Schmidt number is 1, so that the eddy viscosity equals the eddy diffusivity D_t. In the case of a LES model, the diffusion term is calculated using the subgrid-scale viscosity, which represents the advection process by the small-scale turbulent eddies (Appendix A). The contribution of this term is small with respect to the contribution of the large-scale turbulent eddies, which are incorporated in the time-varying velocities.

In the Lagrangian approach, the trajectories of individual particles are calculated. Fluctuations of the particle motions induced by the turbulent motions mimic the mixing processes precisely. The particles are uniformly injected at the inflow cross-section, where the velocity is also uniformly distributed over the cross-section, so that the ensemble of particles is representative for the fluid volume. The particle modelling is described in Chapter 3. The residence time distributions obtained from the ensemble of particles were in correspondence with Eulerian calculations (Chapter 4), indicating that the model gives consistent results.

2.4 PHYSICAL OR CHEMICAL PROCESSES

2.4.1 Ozone decay model

The ozone concentration is calculated by means of an advection-diffusion-decay model, where the ozone consumption is modelled as a first-order decay process. Using the resolved flow field, an ozone concentration (C) can be determined by solving an advection-diffusion-decay model. For an incompressible flow, this equation is given by:

$$\frac{\partial C}{\partial t} + u_i \frac{\partial C}{\partial x_i} = \frac{\partial}{\partial x_i}\left(D_{ij} \frac{\partial C}{\partial x_j}\right) - k_s C, \tag{2.2}$$

where the Einstein summation convention is used (terms with a repeating index are summed up), u_i is the velocity (m/s) in x_i (x, y or z) direction (m), D_{ij} the (eddy) diffusivity coefficient (m^2/s) and k_s the first-order ozone decay (1/s). Because the concentration of micro-organisms is low, it is assumed that the ozone consumption induced by micro-organisms is negligibly small and the ozone is mainly consumed by organic matter. Thus, the calculated particle tracks have no effect on the ozone concentrations. Ozone decay coefficients ranging from 0.001 s^{-1} to 0.01 s^{-1} were used (van der Helm et al., 2007).

2.4.2 UV irradiance model

A multiple segment source summation (MSSS) model was used to determine the UV irradiance (fluence rate) distribution (Bolton and Stefan, 2002; Liu et al., 2004). The UV lamps in the water are schematised as a cylindrical quartz tube containing air and a radiation source in the middle. The radiation source is divided into different segments in longitudinal direction that irradiate UV in all directions. The optical pathways of the radiation are refracted at the air-quartz and quartz-water interface. Snel's law provides the angles of refraction:

$$n_i \sin\theta_i = n_o \sin\theta_o, \tag{2.3}$$

where n_i is the refraction index of the incoming medium, which is crossed by the optical pathway with an angle of θ_i normal to the interface. The index o corresponds the out coming medium. According to Fresnel's law, a part of the radiant energy is reflected at the interface. The reflectance R for unpolarised incident radiant energy is:

$$R = \frac{1}{2}\left(r_{\parallel}^2 + r_{\perp}^2\right), \tag{2.4}$$

where $r_{\perp}$ is the amplitude of the radiant energy perpendicular to the plane of incidence and $r_{\parallel}$ is the amplitude of the radiant energy parallel to the plane of incidence. These amplitudes are calculated by:

$$r_{\parallel} = \frac{n_o \cos\theta_i - n_i \cos\theta_o}{n_i \cos\theta_o + n_o \cos\theta_i},$$

$$r_{\perp} = \frac{n_i \cos\theta_i - n_o \cos\theta_o}{n_i \cos\theta_i + n_o \cos\theta_o}.$$

The UV irradiance is calculated by the inverse square law. For a non-absorbing medium this results in:

$$I_0 = \frac{P}{4\pi l^2}, \tag{2.5}$$

where I_0 is the irradiance without absorption (W/m^2) that depends on the length l of the optical pathway (m) and the power output P of the lamp (W). Absorbance reduces the irradiance, calculated by the Beer-Lambert law:

$$U = T_w(\lambda)^{\frac{l}{l_{10}}}, \tag{2.6}$$

where U is the attenuation factor for the irradiance ($I = I_0 U$), $T_w(\lambda)$ is the transmittance using a 10 mm path length for a given wavelength λ, l is the path length and l_{10} equals 10 mm. We also introduce a specific absorbance (α, [m^{-1}]), which is written as a function of the 10 mm transmittance:

$$\alpha = \frac{-\ln(T_w)}{0.01\ \mathrm{m}}. \tag{2.7}$$

Due to refraction, a focus effect occurs, because the radiant power is concentrated to a smaller cross-section than in the absence of refraction. According to Liu et al. (2004), the focus factor A_f (-) is defined as the cross-sectional area of a circular segment derived from two optical pathways divided by such an area without refraction. The details of the calculation method for the radiation model are explained in Appendix B. The results of the UV irradiance model were compared to measurements of Liu et al. (2004). In these measurements, the irradiance intensity was measured in air. The model showed a good agreement with the measurements (Figure 2.3).

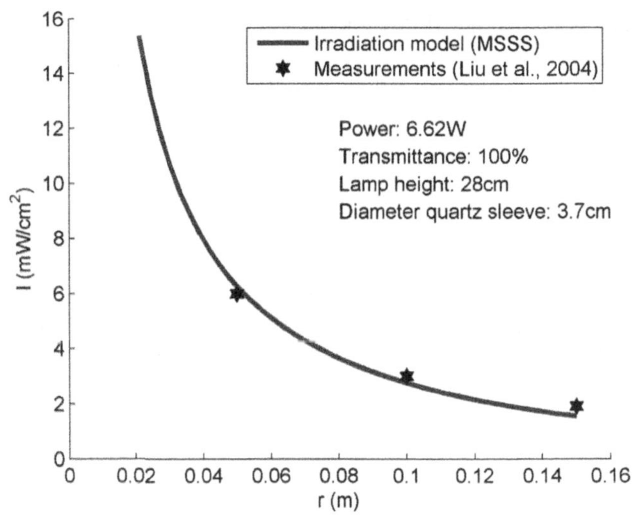

Figure 2.3 Validation of the UV radiation model with measurements in air (Liu et al., 2004).

2.4.3 CT value or UV dose

The particle tracks allow for the calculation of the doses (CT value or UV dose) by integrating the ozone concentration or UV irradiation over each particle path. Using the particles comes closest to the actual movement of micro-organisms, which are also discrete particles. Ducoste et al. (2005) and Sozzi and Taghipour (2006) used this approach to estimate the disinfection performance in UV reactors. By using a large number of particles, important statistics, such as the distribution of doses, are obtained. This distribution provides crucial insight into the hydraulics of the system and provides information for the design and optimisation of ozone contactors and UV systems.

The mean dose can also be calculated by an Eulerian approach, which is referred to as **the Eulerian mean CT value method**. The mean dose is calculated from an advection-diffusion model, where the ozone concentration or UV irradiance is added as a source term (Zhang et al., 2007), similar as in eq. 2.2. Disadvantage of this method is that only a mean value is calculated, whereas higher order statistics are not obtained.

2.5 EFFECTS ON WATER QUALITY

2.5.1 Disinfection model

The simplest disinfection model is the Chick-Watson model, found by Chick (1908) and Watson (1908):

$$\frac{dN}{dt} = -k_{\mu} C^{n} N, \tag{2.8}$$

where N is the concentration of micro-organisms (#/L), n is the coefficient of dilution, k_{μ} is the inactivation rate (in the case of $n = 1$: for ozone [L/(mg·s)], for UV [cm^2/mJ]) and C the disinfectant concentration (for ozone [mg/L], for UV [mW/cm^2]). The integrated form yields:

$$\frac{N}{N_0} = \exp\left(-k_{\mu} C^{n} t\right), \tag{2.9}$$

where N_0 is the initial concentration of micro-organisms. In the case of $n = 1$, the disinfection is proportional to Ct – the disinfectant dose. It is known as CT value or ozone exposure for ozone systems (mg·min/L) and UV dose for UV systems (mJ/cm^2). The Chick-Watson model is often used to predict the inactivation of micro-organisms. A deviation from this relationship may occur at high inactivation levels, the so-called tailing, which means that a smaller increase in inactivation is observed at higher UV doses. Typically, tailing occurs at log inactivation levels of around 4 to 5. Since the extent of removal used in this thesis was lower, no tailing is considered here.

Certain classes of micro-organisms start to be inactivated after a threshold disinfectant dose, which is represented by the delayed Chick-Watson model. The delayed Chick-Watson model after integration yields:

$$\frac{N}{N_0} = \exp\left(-k_{\mu}\left(\phi - \phi_0\right)\right), \tag{2.10}$$

where ϕ represents the disinfectant dose and ϕ_0 represents the threshold value. In Figure 2.4, the log inactivation as a function of CT value or dose is represented for the different variants of the Chick-Watson model.

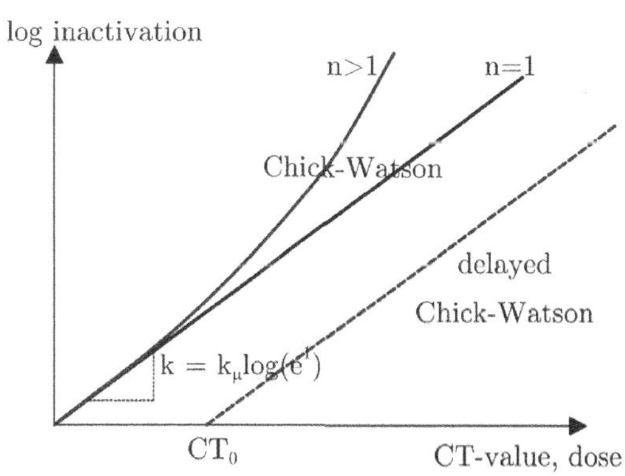

Figure 2.4 Log inactivation as a function of dose described by the Chick-Watson model.

Table 2.1 Overview of inactivation constants for ozone and UV.

Ozone inactivation kinetics				
Reference	**Micro-organism**	$k_{\mu}^{\dagger}$ **(L/(mg·min))**	ϕ_0 **(mg·min/L)**	**Temp. (°C)**
Huang et al. (2004)	*Cryptosporidium*	0.441	0	–
USEPA (2006a)	*Cryptosporidium*	0.24	0	10
Craik (2005)	*Cryptosporidium*	0.852	0	22
Rennecker et al. (1999)	*Cryptosporidium*	0.8	0.74	20
Zhou and Smith (1994)	*Escherichia coli*	499	0	10
von Gunten (2003a)	*Escherichia coli*	130	0	20
von Gunten (2003a)	*Bacillus subtilis*	2.9	2.9	20
AWWA (1991)	*Giardia lamblia*	4.9	0	10
UV inactivation kinetics				
Reference	**Micro-organism**	$k_{\mu}^{\dagger}$ **(cm^2/mJ)**	ϕ_0 **(mJ/cm^2)**	DE_{max}
Hijnen et al. (2006)	*Bacillus subtilis*	0.132	12.3	4.0
Hijnen et al. (2006)	*Escherichia coli*	1.17	0	6.0
Hijnen et al. (2006)	*Clostridium p.*	0.138	18	3.0
Hijnen et al. (2006)	MS2 phage	0.127	0	4.9
Hijnen et al. (2006)	Qβ	0.193	0	4.2
Hijnen et al. (2006)	T7	0.534	0	4.6

†In literature, the inactivation is sometimes represented as: $^{10}\log\left(\frac{N}{N_0}\right) = -k\phi$, for which the inactivation constant k differs with a factor of $^{10}\log(e^1)$ from the inactivation constant k_{μ} used here.

The inactivation constant is determined from batch experiments for ozone processes and collimated beam experiments for UV processes. An overview of inactivation constants for different micro-organisms is given in Table 2.1. The value for DE_{max} represents the maximal log removal after which tailing occurs. The inactivation constants for disinfection with ozone are temperature dependent, which can be described by Arrhenius law.

The inactivation of micro-organisms can be determined from the doses calculated for the particles. This method is denoted as **particle tracking method**. Summation over all the particles gives the total inactivation, shown as the fraction of surviving micro-organisms from the total amount of micro-organisms:

$$\frac{N}{N_0} = \frac{1}{N_0}\sum_{i=1}^{N}\exp\left(-k_{\mu}\left(\phi_i - \phi_0\right)\right), \tag{2.11}$$

where ϕ_i represents the disinfectant dose of particle i. The total inactivation is usually presented as a decimal elimination (DE), so that the log inactivation yields:

$$DE = -^{10}\log\left(\frac{N}{N_0}\right). \tag{2.12}$$

In the **Eulerian direct inactivation method**, the inactivation is determined directly by solving the transport equation for a concentration of micro-organisms N (Huang et al., 2004; Greene et al., 2006). The ozone concentration or UV irradiance multiplied by the inactivation rate constant appears as a source term in the advection-diffusion equation (2.2). In absence of a velocity and (turbulent) diffusion, the equation reduces to the classical Chick-Watson equation.

2.5.2 Advanced oxidation model

For the advanced oxidation process, we consider UV in combination with hydrogen peroxide. Organic compounds can be degraded by the effect of direct photolysis and/or the reaction with hydroxyl ($OH\bullet$) radicals. Hydroxyl radicals are formed by the photolysis of hydrogen peroxide (H_2O_2). The reaction by photolysis for a certain compound N_i (organic compound or hydrogen peroxide for example, mol/L) is given by (Sharpless and Linden, 2003):

$$\frac{d[N_i]}{dt} = -E_{\text{CFD}}\frac{1-10^{-A}}{A}\Phi_{N_i}\varepsilon_{N_i}[N_i], \tag{2.13}$$

where Φ represents the quantum yield, ε the molar extinction (L/(mol·cm)) and A represents the absorbance, which can be calculated from the summation of the molar extinction times the concentration over all the compounds in the water ($A = l\sum \varepsilon_{N_i}[N_i]$) or from the 1 cm transmittance (T_w) of the water ($A = -{}^{10}\log(T_w)$). E_{CFD} is the photon flux (mmol/(cm^2·s)), for which a unit conversion is needed from the UV intensity I (mW/cm^2) as calculated by the CFD model:

$$E_{\text{CFD}} = \frac{I}{N_A E_f}, \tag{2.14}$$

where E_f is the energy of a photon (J) and N_A represents Avogadro's number (1/mol). For the degradation of organic compounds, next to direct photolysis, also a reaction with $OH\bullet$ radicals occurs, so that the equation for the total reaction of compound N_i becomes:

$$\frac{d[N_i]}{dt} = -E_{\text{CFD}}\frac{1-10^{-A}}{A}\Phi_{N_i}\varepsilon_{N_i}[N_i] - k_i[OH\bullet][N_i], \tag{2.15}$$

where k_i represents the reaction rate of contaminant i with $OH\bullet$ radicals (L/mol/s) and $[N_i]$ the concentration of contaminant i (mol/L). The $OH\bullet$ radicals are formed by photolysis of hydrogen peroxide, whereas the $OH\bullet$ radicals react with the different compounds (i) in the water:

$$\frac{d[OH\bullet]}{dt} = 2E_{\text{CFD}}\frac{1-10^{-A}}{A}\Phi_{H_2O_2}\varepsilon_{H_2O_2}[H_2O_2] - \sum k_i[N_i][OH\bullet]. \tag{2.16}$$

The photolysis of hydrogen peroxide results in the formation of two $OH\bullet$ radicals, which explains the factor two. For the reaction of $OH\bullet$ radicals with compound i it is assumed that one radical reacts with one compound molecule. The hydroxyl radicals react very quickly with different contaminants in the water so that a steady-state concentration of $OH\bullet$ radicals immediately forms (de Laat et al., 1999; Sharpless and Linden, 2003). The steady-state concentration for the $OH\bullet$ radicals is then calculated by:

$$[OH\bullet] = 2E_{\text{CFD}}\frac{1-10^{-A}}{A}\frac{\Phi_{H_2O_2}\varepsilon_{H_2O_2}[H_2O_2]}{\sum k_i[N_i]}. \tag{2.17}$$

The factor $\sum k_i[N_i]$ contains the reactions with all the compounds (including peroxide) in the water. Since the number of compounds can be very large, the reactions with the background components in the water were not individually solved for all the background components but treated as one overall reaction. This factor should be determined by collimated beam experiments for each type of water. By substituting the steady-state $OH\bullet$ radical concentration into eq. 2.15, the conversion of a contaminant N_0 by direct photolysis and reaction with $OH\bullet$ radicals can be written as:

$$\frac{d[N_0]}{dt} = -E_{\text{CFD}}\frac{1-10^{-A}}{A}\left(\Phi_{N_0}\varepsilon_{N_0} + 2\frac{\Phi_{H_2O_2}\varepsilon_{H_2O_2}[H_2O_2]}{\sum k_i[N_i]}k_0\right)[N_0], \qquad (2.18)$$

where k_0 represents the reaction rate of contaminant N_0 with $OH\bullet$ radicals. This equation represents a (pseudo-)first-order reaction.

The hydrogen peroxide concentration can be calculated by solving eq. 2.18 (by using H_2O_2 instead of compound N_0). In total a system of two differential equations needs to be solved, for hydrogen peroxide and the compound of interest. However, because the reduction of hydrogen peroxide is small, we could also assume that it remains constant in the calculation. Both approaches were explored in Chapter 5.

The oxidation can be calculated in an Eulerian framework, solving concentrations of peroxide and organic substances, or in a Lagrangian framework. In the latter, from the dose calculated for each particle, a UV irradiance and irradiation time was chosen over which the differential equation for the compound removal (eq. 2.18) was numerically integrated, resulting in a degradation for each particle. Averaging over all the particles resulted in the overall degradation by advanced oxidation. In Chapter 5 it is shown that both approaches give similar results. The compound atrazine is considered in this thesis. In Table 2.2, the reaction constants for peroxide and atrazine found in literature are shown.

Table 2.2 Overview of reaction constants for the UV/H_2O_2 process.

Reaction H_2O_2 with $OH\bullet$	$k_{H_2O_2}$†	2.7×10^7	de Laat et al. (1999)
Reaction atrazine with $OH\bullet$	k_{ATZ}†	2.6×10^9	Parsons (2004)
Quantum yield H_2O_2	$\Phi_{H_2O_2}$	0.5	de Laat et al. (1999)
Quantum yield atrazine	Φ_{ATZ}	0.05	Nick et al. (1992)
Molar extinction H_2O_2	$\varepsilon_{H_2O_2}$*	18.6	de Laat et al. (1999)
Molar extinction atrazine	ε_{ATZ}*	3860	Nick et al. (1992)

†L/mol/s, *L/mol/cm

Diffusion limitation

Diffusion limitation occurs when reaction rates are so high that the supply of reactants is limited by diffusion. The question is whether this effect is occurring during the UV advanced oxidation process. Therefore, we consider a small element of water that is irradiated by UV, and estimate the reaction time scales. The following reactants are considered:

- $OH\bullet$ radicals are produced when the hydrogen peroxide is irradiated by UV. The $OH\bullet$ radicals react with substances in the water (hydrogen peroxide, atrazine, etc.).
- Hydrogen peroxide concentration is lowered due to direct photolysis and reactions of hydrogen peroxide with $OH\bullet$ radicals.
- Contaminant concentration, such as atrazine, is lowered reduced due to direct photolysis and reactions with $OH\bullet$ radicals.

The concentrations of these substances can be determined from the rate equations described above. In Figure 2.5 on the left panel, the normalised concentration of these three reactants are shown as a function of time. This example considered the degradation of atrazine under realistic conditions (kinetic constants were used from Table 2.2). The instantaneous $OH\bullet$ radical concentration was obtained by solving eq. 2.16, whereas the concentration of peroxide as well as atrazine was computed by solving eq. 2.15 for each compound using the instantaneous concentration of $OH\bullet$ radicals. In less than $1 \cdot 10^{-4}$ s, the $OH\bullet$ radical concentration was constant, which was equal to the steady-state concentration given by eq. 2.17. No significant amount of peroxide or atrazine had reacted in this short time, confirming that it is permitted to use the steady-state $OH\bullet$ radical concentration. Significant amounts of atrazine were degraded after $1 \cdot 10^{-2}$ s, whereas for the hydrogen peroxide these time scales were larger. On the right panel of Figure 2.5, the half-lives are shown as a function of irradiation. By increasing the irradiation, more $OH\bullet$ radicals were formed, so the atrazine and peroxide were degraded faster. At the highest irradiation level the time scales were still larger than the turbulent diffusion time scales (Kolmogorov time scales $\sim 10^{-3}$ s), so that diffusion limitation did not play a role for the advanced oxidation of atrazine. In other words, before the hydrogen peroxide and/or contaminant is substantially degraded, hydraulic processes provide fresh concentrations to the reaction zone, so that there are sufficient reactants available for the $OH\bullet$ radicals to react with.

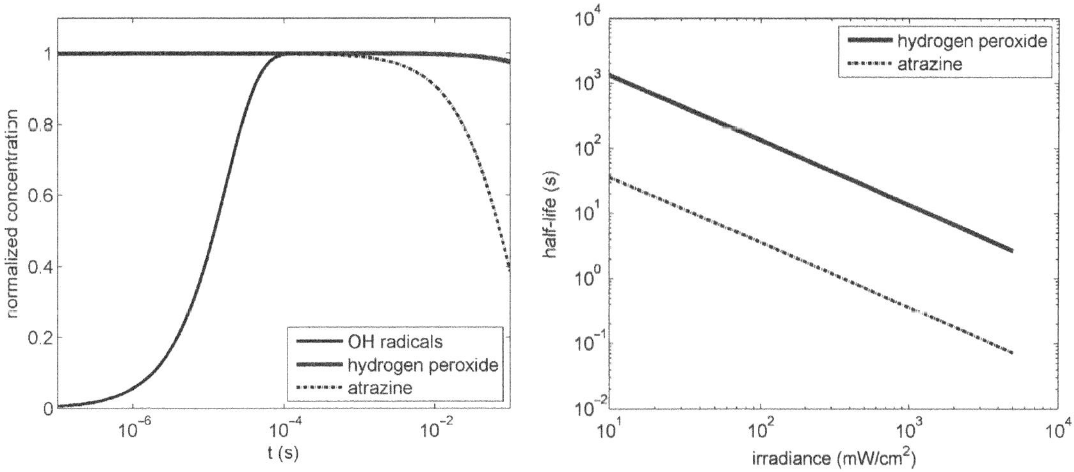

Figure 2.5 Left panel: Normalised concentrations of the reactants for a water element irradiated by UV (5000 mW/cm^2). Right panel: The half-life for hydrogen peroxide and atrazine in a water element as a function of the irradiance. Initial concentration peroxide: 10 mg/L, initial concentration atrazine: 2 μg/L, background reaction rate: $\sum k_i N_i = 5 \cdot 10^4$ s^{-1}.

2.6 ALTERNATIVE DISINFECTION MODELS FOR OZONE SYSTEMS

The accuracy of a disinfection prediction increases by considering an increasing number of calculated or measured variables, however, at a cost of higher computational resources. The CFD model with a particle tracking method accounts for most of the required variables (movement of particles and distribution of ozone concentration), which results in a distribution of doses. It comes closest to the actual movements of individual micro-organisms in reality, which is expected to result in the most accurate disinfection prediction. Apart from the CFD modelling approaches, more simple methods exist in practise to predict the disinfection at much lower computational costs. For ozone systems, these methods are: the CSTR model, CT_{10} method, segregated flow analysis and micro mixing analysis. In "the surface water treatment rule" (USEPA, 1991), the CT_{10} method is used to determine the disinfection from CT tables, and in "the long term 2 enhanced surface water treatment rule" (USEPA, 2006b), both the CSTR method and CT_{10} method are used. The above mentioned methods will be explained in this section, and will be assessed in Chapter 4 for different ozone contactor designs. Next to these methods, an analytical expression for the maximum disinfection capacity is proposed, which shows the potential improvement of an ozone system.

2.6.1 Maximum disinfection capacity

From the kinetic constants (inactivation rate constant and ozone decay coefficient) and the hydraulic residence time, a maximal removal in the case of perfect hydraulics can be formulated for the ozone system. This is represented by a perfect plug flow system, where every fluid element receives an equal amount of ozone. For a first-order ozone decay, the ozone concentration as a function of time is substituted into eq. 2.8 (using $n = 1$). After division by N, integration from 0 to the mean residence time T_h and substituting the result into eq. 2.12, the maximal log removal becomes ($DE_{\max}$):

$$DE_{\max} = {}^{10}\log(\mathrm{e})\frac{k_{\mu} C_0}{k_s}\left(1 - \exp(-k_s T_h)\right), \tag{2.19}$$

where T_h is the hydraulic residence time (s), which may reduce the maximum log removal in the case that not all the ozone is consumed in the system.

2.6.2 CSTR method

In the CSTR method, every compartment inside the ozone contactor is regarded as a CSTR (continuously stirred tank reactor), for which two mass balances are formulated (Levenspiel, 1972). A mass balance for the disinfectant concentration and a mass balance for the concentration of micro-organisms are solved in every CSTR to determine the disinfection capacity. For a number of m CSTRs-in-series, the disinfectant concentration in CSTR j reads:

$$\frac{C_j}{C_0} = \frac{1}{\left(1 + \frac{T_h}{m} k_s\right)^j}, \tag{2.20}$$

where C_j is the disinfectant concentration in the jth CSTR (mg/L), C_0 the initial concentration of disinfectant (mg/L). From the mass balance of the micro-organisms, the total inactivation after n CSTRs-in-series is calculated (Smeets et al., 2006):

$$\frac{N}{N_0} = \prod_{j=1}^{m} \frac{1}{1 + \frac{T_h}{m} k_{\mu} C_j} \tag{2.21}$$

No residence time distributions are needed to calculate the disinfection, although they can be used as a fit for the number of CSTRs. Variations on this approach exist that include the modelling of dead zones or short-circuits to reproduce more details of the complex hydraulics, described by Martin-Dominguez et al. (2005).

2.6.3 CT_{10} method

In the surface water treatment rule guidance manual (USEPA, 1991), the CT concept is applied, where the microbial inactivation is calculated from the product of the residual ozone outlet concentration and a characteristic dose. For the CT_{10} method, the ozone concentrations at different positions in the installation are taken, which are multiplied by the T_{10}, the time when 10% of the fluid elements have passed that position. The CT_{10} is usually calculated at the outlet of each compartment. The inactivation is then calculated from a summation of the CT_{10} values at each compartment:

$$\frac{N}{N_0} = \exp\left(-k_{\mu}\sum_{j=1}^{m} C_j T_{10}^j\right), \tag{2.22}$$

where C_j is the ozone concentration (mg/L), T_{10}^j the T_{10} (s) in compartment j and m is the number of compartments, over which the CT_{10} is determined. The T_{10} can be estimated in different ways, depending on the information available. For a rough approximation, a rule of thumb is used that defines the T_{10} as 60% of the hydraulic residence time. Tracer measurements or CFD calculations can be used to determine the T_{10}. The ozone concentrations in the contactor can be determined in different ways: from measurements, calculated by a first-order decay (in combination with the mean residence time) or calculated from the CFD calculations. Furthermore, the ozone concentration of each compartment can be chosen in different ways: the effluent ozone concentration of a compartment or a concentration averaged over the inlet and outlet of a compartment (Rakness et al. (2005)). In this thesis, the concentrations and T_{10} were calculated from the CFD model, where the concentrations were averaged over the inlet and the outlet of each compartment.

2.6.4 Segregated flow analysis (SFA) method

The segregated flow analysis is incorporated in the integrated disinfection design framework, which applies an integrated approach to determine the microbial inactivation (Ducoste et al., 2001). The fluid that enters the installation is divided into discrete elements, which are large with respect to the molecules and small with respect to the total volume of the installation. There is no mixing or transfer of material between these elements; they remain completely segregated. Every fluid element has a certain residence time, determined from tracer tests or CFD calculations. The ozone concentration for each fluid element is also determined from the residence time, assuming a first-order decay kinetics. From this, the CT value $\phi(t)$ is calculated for each fluid element with residence time t:

$$\phi(t) = \int_0^t C(t)dt = \frac{C_0}{k_s}\left(1 - \exp(-k_s t)\right). \tag{2.23}$$

Combining the first-order ozone decay with the Chick-Watson equation, the SFA calculates the inactivation as a function of residence time in the following manner:

$$\frac{N(t)}{N_0} = \exp\left(\frac{k_{\mu}C_0}{k_s}\left(1 - \exp(-k_s t)\right)\right). \tag{2.24}$$

The total inactivation then follows from:

$$\frac{N}{N_0} = \int_0^{\infty} \frac{N(t)}{N_0} f(t)dt, \tag{2.25}$$

where $f(t)$ is the residence time distribution. So, the SFA method only needs a residence time distribution as an input variable.

2.6.5 Micro-mixing analysis (MMA) method

In contrast with the SFA method, where the flow segments are assumed to remain completely segregated, the micro-mixing analysis assumes that elements of fluid mix completely and transfer material between elements on a molecular scale. Instead of the residence time, the life expectancy λ is used, which is defined as the expected time that a fluid element will remain in the installation. The life expectancy density distribution function $f(\lambda)$ is related to the residence time distribution function $f(t)$, see Zwietering (1959). The following differential equations describe the ozone concentration and concentration of micro-organisms (Craik, 2005):

$$\frac{dC}{d\lambda} = -k_s C + (C - C_0)\frac{f(\lambda)}{1 - F(\lambda)}, \tag{2.26}$$

$$\frac{dN}{d\lambda} = -k_\mu CN + (N - N_0)\frac{f(\lambda)}{1 - F(\lambda)}, \tag{2.27}$$

where $F(\lambda)$ is the cumulative life expectancy distribution function. The number of micro-organisms at the outlet was obtained by solving eq. 2.26 and 2.27 at $\lambda = 0$ s, which is an initial value problem with a negative time step. The boundary condition at $\lambda = \infty$ for these equations are $\frac{dC}{d\lambda} = 0$ and $\frac{dN}{d\lambda} = 0$, respectively. According to Zwietering (1959), it is sufficient to start at $\lambda = 4T_m$, where the derivatives are approximately zero. The values of C and N at $\lambda = 4T_m$ are determined iteratively using the method of bisection with initial boundaries for both variables of 0.9999 and 0.0001.

2.7 UV PARAMETER STUDY

For the design and operation of UV systems some useful parameters are introduced here, which were derived from simple mathematical models. A simple expression for the mean dose is proposed, which estimates the mean dose for an arbitrary system with a certain lamp power, flow rate and UV absorption in the water. This expression was also used to normalise the calculated dose from the CFD model. Another important design parameter for UV systems is the distance between the lamps (quartz sleeve) and outer wall of the reactor. An expression for this distance depending on the transmittance of the water is proposed. Lastly, the dose distribution is approximated by an analytical probability density function. This approach allows to make simple estimates of the amount of energy that can be saved by narrowing the dose distribution. In Chapter 6 it is shown that these estimates were comparable to the results of actual dose distributions calculated by the CFD model for different types of UV reactors.

2.7.1 Scale factor mean dose

A scale factor for the mean dose is introduced, which gives an indication of the average dose that could be obtained for a UV reactor with certain operating conditions, such as lamp power, flow rate and UV absorption in the water. The calculation was based upon an annular system (Figure 2.6). We define the lamp radius r_{lamp} as the distance from the centre of the lamp to the quartz sleeve (m) and R as the distance from the quartz sleeve to the outer wall of the reactor (m). A fixed flow rate Q is fed into the system (m^3/s), so that the mean velocity (u_m, [m/s]) yields:

$$u_m = \frac{Q}{\pi(2r_{\text{lamp}}R + R^2)}. \tag{2.28}$$

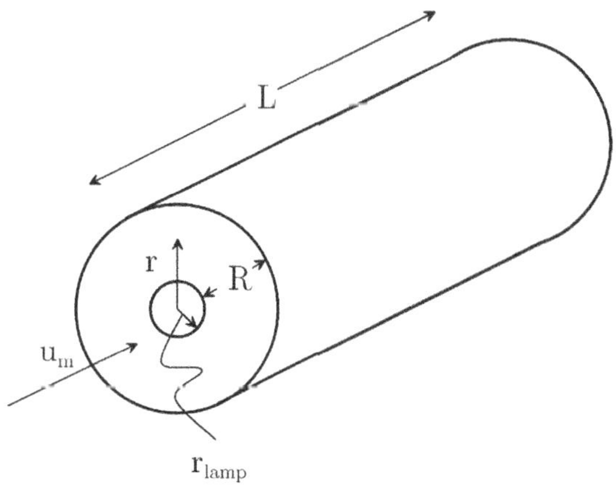

Figure 2.6 Sketch of the annular system used for the calculation of the characteristic dose and for the estimation of the optimal distance from the quartz sleeve to the outer wall (*R*).

The irradiance field I (W/m^2) as a function of radius r is given by (for a 2D system):

$$I(r) = \frac{P}{2\pi Lr}\exp\left(-\alpha(r - r_{\text{lamp}})\right), \tag{2.29}$$

where P is the total power of the lamp (W) and α is the specific absorbance of the water (1/m), given by eq. 2.7. Assuming a constant velocity profile (with velocities equal to u_m), the dose (D, [J/m^2]) as a function of r is calculated by:

$$D(r) = \frac{I(r)L}{u_m} = \frac{PV}{2\pi QL}\frac{\exp\left(-\alpha(r - r_{\text{lamp}})\right)}{r}, \tag{2.30}$$

where V is the total volume of the reactor (m^3), given by $V = L\pi(2r_{\text{lamp}}R + R^2)$, L is the arc length of the lamp (m). By assuming a radius of the reactor equal to $R = 2/\alpha$, at which distance 86% of the radiation is absorbed, the volume of the reactor reduces to $V = \frac{4\pi L}{\alpha^2}(\alpha r_{\text{lamp}} + 1)$. The characteristic dose (D_{scale}, [J/m^2]) is then determined by calculating the dose at $r = 1/\alpha + r_{\text{lamp}}$, which results in a simple analytical expression:

$$D_{\text{scale}} = \frac{2P}{Q\alpha}\exp(-1), \tag{2.31}$$

containing the lamp power, flow rate and UV absorption in water.

2.7.2 Optimal distance from the lamp to the outer wall

The distance from the lamp to the outer wall can be optimised using the analytical model described in §2.7.1. On the one hand, the distance R from the quartz layer of the lamp to the outer wall must not be too large to prevent areas with small doses due to the small UV intensities near the wall. On the other hand, this distance must not be too small to avoid the loss of radiant energy through the outer walls. Therefore, by substituting the dose distribution of eq. 2.30 into a first-order Chick-Watson relation, the disinfection is calculated as a function of the parameter $R\alpha$, which contains the distance R and the UV absorbance in water α. Increasing the parameter $R\alpha$ results in a higher mean dose, because less UV radiation is lost through the outer walls. However, the minimal dose has decreased because the water elements close to the outer wall received less UV radiation (Figure 2.7, left panel). The log removal as a function of $R\alpha$ depends on the inactivation rate k_μ, lamp power P_0 and flow rate Q (Figure 2.7, right panel). The optimal value of $R\alpha$ depends on the required removal of contaminants. For a required log removal between 1 and 3, the optimal distance R is roughly around $1/\alpha$ (which is $-0.01/^{10}\log(T_w)$). For a T_w transmittance of 80%, the optimal distance then becomes approximately 0.05 m.

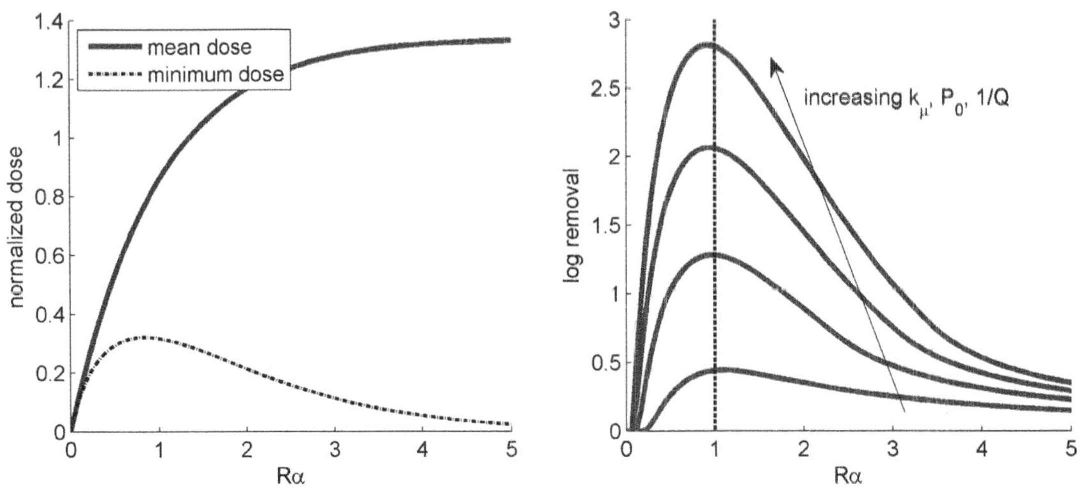

Figure 2.7 Normalised mean dose and minimal dose as a function of the parameter $R\alpha$ (left panel). Log removal as a function of the parameter $R\alpha$ (right panel).

2.7.3 Approximation of dose distribution

The dose distribution delivered by a UV system can also be approximated by a probability density function (pdf). This forms an easy tool to perform a sensitivity analysis: The effect of narrowing the dose distribution on the energy consumption can then be explored for different kinetic constants. A realistic dose distribution is given by the (shifted) exponential dose distribution, which is given by:

$$f(D) = \lambda \exp\left(-\lambda(D - D_{\min})\right). \tag{2.32}$$

The mean dose then yields:

$$D_{\text{mean}} = \frac{1}{\lambda} + D_{\min} \tag{2.33}$$

The dose distribution is characterised by the factor D_{10}/D_{mean} (D_{10} is the 10th percentile of the dose). When D_{10}/D_{mean} equals 1, the system resembles a perfect plug flow with no variations in doses. If D_{10}/D_{mean} is smaller than 1, the hydraulics of the system are suboptimal. For a range of D_{10}/D_{mean}, the dose distributions were simulated according to eq. 2.32, for which the energy consumption corresponding to a certain disinfection or oxidation level was calculated. This was calculated as follows: for each value of the factor D_{10}/D_{mean}, a mean dose was calculated that corresponds to the required log removal or log degradation. Should the distribution become wider, a higher mean dose was needed to obtain a similar removal or degradation. It was assumed that the power of the UV lamps scales linearly with the mean dose, so that the energy consumption as a function of the dose distribution (factor D_{10}/D_{mean}) was determined for each required extent of removal or degradation. This resulted in the differences in energy consumption as a function of the factor D_{10}/D_{mean}. The calculated energy (E) was normalised with the energy of a perfect plug flow (E_{ideal}), which is defined as the energy consumption at $D_{10}/D_{mean} = 1$. The maximum energy reduction was obtained when the dose distribution is reduced this value. We could for example assume that optimising the hydraulics of a UV reactor by modifying the geometry may result in a narrowing of the dose distribution from $D_{10}/D_{mean} = 0.5$ to $D_{10}/D_{mean} = 0.75$. This optimisation results in a reduction of energy, which is represented as a percentage of the energy consumption before the optimisation (Figure 2.8). Three parameters are distinguished that influence the energy reduction that is obtained by narrowing the dose distribution:

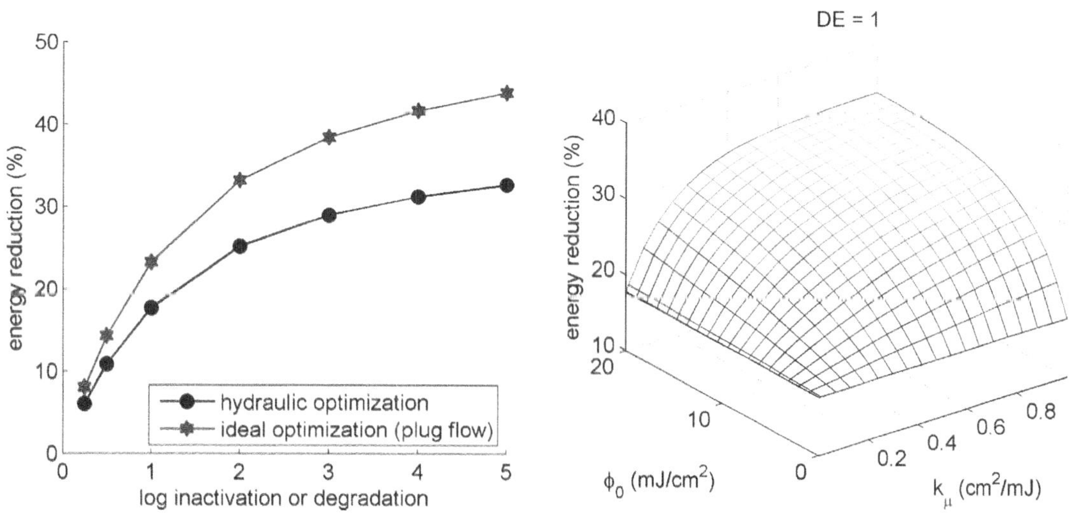

Figure 2.8 Energy reduction as a function of required log inactivation or degradation (left panel), decay constant and threshold dose (right panel) in the case of the dose distribution being narrowed from $D_{10} = 0.5D_{mean}$ to $D_{10} = 0.75D_{mean}$.

- The total extent of removal. In Figure 2.8 on the left, the energy reduction that occurs by optimising the hydraulics is plotted as a function of the required log inactivation or degradation. At a higher required extent of removal, a higher percentage in energy reduction can be obtained because the differences in doses have a greater effect. Since the energy consumption is also higher at a higher log removal, the absolute reduction in energy due to the hydraulic optimisation will even be higher.

- The first-order decay constant, which can be derived from eq. 2.10 for disinfection and eq. 2.18 for oxidation. In Figure 2.8 on the right, the energy reduction is plotted as function of the first-order decay and threshold dose for a total removal of log 1. For this fixed extent of removal, the energy reduction does not depend on the decay coefficient when the threshold dose is equal to zero. But in the case of a threshold dose larger than zero, the energy reduction increases at higher decay constants. Since lower doses are needed at higher decay rates to obtain the required amount of removal, the threshold dose has a bigger impact, so that narrowing the dose distribution has a larger effect.
- The threshold dose. At higher threshold doses, the energy reduction obtained from optimisations increases (Figure 2.8 on the right) because the lower doses will even be more critical for disinfection, so that a reduction of these low doses is more effective.

It is concluded that optimising the flow may result in energy reductions ranging from 25% up to 50% for an ideal hydraulic behaviour. For a drinking water treatment plant, optimising the flow of an UV installation may result in a considerable reduction in costs. For example, a plant with an average flow rate of $Q = 10\,000$ m^3/h requires a power output for UV disinfection of 500 kWh, assuming a power use of 50 Wh/m^3 for disinfection (de Moel et al., 2004). By optimising the flow, the energy is reduced by 25%, which results in a yearly cost reduction of 90 thousand euros (for an energy price of 0.08 eurocent per kWh).

Chapter 3

Particle tracking*

In the previous chapter we have seen that the particle tracking routine forms an essential part of the CFD modelling to predict the performance of treatment installations. We will focus on the particle modelling in this chapter, which involves a mathematical description of the particle model and the numerical implementation.

3.1 INTRODUCTION

For predicting disinfection processes, transport and mixing mechanisms of chemical and microbial species are important. The combination of CFD with particle tracking is therefore a useful tool, since it captures the complex flow features and results in a distribution of doses. The particles might represents individual micro-organisms, for which the disinfection performance can be estimated accurately. The main advantage of a particle tracking (Lagrangian) method over an Eulerian method is that different quantities (e.g., residence time, ozone exposure) can be recorded for every particle, resulting in higher-order statistics for the desired quantity. The accuracy of a particle tracking method, however, is a point of concern (Zhang et al., 2007). A thorough examination of different particle tracking methods is therefore conducted and specific problems are addressed. For example, special attention is needed to prevent particles from, unrealistically, accumulating at regions with low diffusion and to prevent them from virtually crossing the walls, which is in practice not possible of course. The particles were assumed to represent micro-organisms or micropollutants that move entirely with the fluid (particles without gravity, drag or lift forces). This approach is allowed if particles have a density close to the density of the fluid, and if they are smaller than the Kolmogorov length scales ($\sim 1 \cdot 10^{-4}$ m), so that the velocity of the particle is the same as the fluid velocity (Baldyga and Orciuch, 2001), which is valid for micro-organisms in water (for example, one of the larger micro-organisms, *Bacillus subtilis* spores, has a length of 1–5 μm).

The theory of stochastic differential equations was used in §3.2 to find the appropriate particle displacement equations that represent the advection-diffusion process. The particle displacement was divided into an advection part and a diffusion part. For the k-ε model, both parts need to be taken into account, whereas for the LES model, the advection part suffices. For the advection part in §3.3, a semi-analytical

* Parts of the research described in this chapter are published in: Wols, B.A., Hofman, J.A.M.H, Uijttewaal, W.S.J., Rietveld, L.C., Stelling, G.S. and van Dijk, J.C., A particle tracking technique to estimate disinfection efficacy in drinking water treatment plants. In *6th International Conference on CFD in Oil & Gas, Metallurgical and Process Industries*, Trondheim, 2008.

scheme (Dunsbergen, 1994) was used. From the theory of stochastic differential equations, numerical schemes were explored in §3.4 that modelled the diffusion of particles by random displacements accurately. The particle tracking scheme was tested for a channel flow in §3.5. Finally, some remarks are made in §3.6 with respect to the number of particles required for an accurate disinfection prediction.

3.2 STOCHASTIC DIFFERENTIAL EQUATIONS

3.2.1 Brownian motion

The molecular motion of small-sized bodies, suspended in a liquid, is called Brownian motion, discovered by Brown (1829), and is further explained by Einstein (1905). Langevin and others tried to formulate the Brownian motion process in terms of differential equations, written as:

$$dX_t = f(t, X_t)dt + g(t, X_t)dW_t, \tag{3.1}$$

$$X_t(0) = X_0, \tag{3.2}$$

where X_t is a stochastic variable and dW_t represents a random increment. The function $f(t, X_t)$ represents a drift term and $g(t, X_t)$ a noise term. This equation is known as a stochastic differential equation (sde) and requires a different approach than deterministic differential equations (Risken, 1984). The increments dW_t are usually Gaussian with zero mean and a variance equal to dt. The increments can be generated by a Gaussian random number generator (with mean zero and variance 1):

$$\Delta W_t = N(0,1)\sqrt{\Delta t}, \tag{3.3}$$

so the increment is a normal distributed random variable with mean zero and variance Δt. Instead of using a Gaussian distribution, which resulted in a Gaussian white noise process, we also investigated the use of a uniform distribution, which resulted in a uniform white noise process. The increments were then generated by a uniform random number generator (with mean zero and variance 1).

3.2.2 Itô vs Stratonovitch

The stochastic differential equation must be interpreted as a stochastic integral equation in order to have a meaning. The integral form of sde 3.1 yields:

$$X_t = X_0 + \int_{t_0}^{t} f(s, X_s)ds + \int_{t_0}^{t} g(s, X_s)dW_s, \quad 0 \le t_0 \le t \le \infty. \tag{3.4}$$

The last term on the right hand side is a stochastic integral, which can not be defined as a classical Riemann-Stieltjes integral due to the random variable. We need to interpret this stochastic integral to give the equation a useful meaning. Well-known interpretations are the Itô interpretation and Stratonovitch interpretation (Kloeden and Platen, 1992), depending on the time at which the function g is evaluated. The Itô stochastic integral is defined in the following mean square limit (Stijnen, 2002):

$$\int_{t_0}^{t} g(s, X_s)dW_s = \lim_{\Delta t \to 0} \sum g(t_i, x)\left(W_{t_{i+1}} - W_{t_i}\right), \tag{3.5}$$

where the function g is evaluated at the beginning of the interval t_i to t_{i+1}. The Stratonovitch interpretation results in the following mean square limit (Stijnen, 2002):

$$\int_{t_0}^{t} g(s, X_s)dW_s = \lim_{\Delta t \to 0} \sum g\left(\frac{t_{i+1}+t_i}{2}, x\right)\left(W_{t_{i+1}} - W_{t_i}\right), \tag{3.6}$$

where the function g is evaluated at the middle of the interval. For a proper definition of a stochastic differential equation, the integration rules need to be defined, otherwise the sde is meaningless. For the same process, an Itô sde as well as a Stratonovitch sde can be defined, which are equivalent, but have a different appearance. The Itô sde of the Brownian motion process reads:

$$dX_t = f(t, X_t)dt + g(t, X_t)dW_t, \tag{3.7}$$

whereas the Stratonovitch sde for the same process reads:

$$dX_t = f(t, X_t)dt + g(t, X_t)dW_t - \frac{1}{2}g(t, X_t)\frac{\partial g}{\partial x}|_{(t, X_t)}dt. \tag{3.8}$$

From these definitions a transformation rule between both the interpretations is obtained (van Kämpen, 1981). In the case g only depends on time, both the formulations coincide. The Itô interpretation is most often used, and is used here, because it provides the simplest way of calculation.

3.2.3 Fokker-Planck equation

The stochastic differential equations can be related to an Eulerian process by considering the probability density function $p(x, t|x_0, t_0)$, which gives the transition probability $p(x, t|x_0, t_0)dx$ of X_t from the value x_0 at t_0 into the interval x, $x + dx$ at t. The probability density function can be obtained by solving the Fokker-Planck equation. For the Itô sde, an Itô-Fokker-Planck equation can be constructed (Risken, 1984):

$$\frac{\partial p}{\partial t} = -\frac{\partial}{\partial x}\left(f(t, x)p\right) + \frac{1}{2}\frac{\partial^2}{\partial x^2}\left(g^2(t, x)p\right), \tag{3.9}$$

with initial condition $p(x, t_0|x_0, t_0) = \delta(x - x_0)$. The Fokker-Planck equation is a deterministic partial differential equation. In this way, a sde can be compared to a diffusion or an advection-diffusion process and the functions f and g can be defined that correspond to the actual process. We will evaluate these functions for a pure diffusion (Fickian) process and an advection-diffusion process.

3.2.4 Fickian diffusion process

In the case of a Fickian diffusion process, the differential equation for a concentration C yields:

$$\frac{\partial C}{\partial t} = \frac{\partial}{\partial x}D\frac{\partial C}{\partial x}. \tag{3.10}$$

where D is a non-uniform diffusion coefficient, depending on x. By substituting $p = C$ into the Fokker-Planck equation 3.9, the drift function $f(t, x)$ and noise function $g(t, x)$ can be determined that correspond to the Fickian diffusion process:

$$f(t,X_t) = \frac{dD}{dx}, \tag{3.11}$$

$$g(t,X_t) = \sqrt{2D}, \tag{3.12}$$

where $f(t,X_t)$ represents the so-called drift term, which prevents particles from accumulating at regions with low diffusion and $g(t,X_t)$ represent the noise term, which is responsible for the random displacement. The underlying sde that describes a Fickian diffusion process is then equal to:

$$dX_t = \frac{dD}{dx}dt + \sqrt{2D}dW_t. \tag{3.13}$$

3.2.5 Advection-diffusion process

The particle tracking routine used in this thesis must obey the advection-diffusion equation, which is given by (in multiple dimensions):

$$\frac{\partial C}{\partial t} + u_i\frac{\partial C}{\partial x_i} = \frac{\partial}{\partial x_i}\left(D_{ij}\frac{\partial C}{\partial x_j}\right), \tag{3.14}$$

using the Einstein summation convention (terms with a repeating index are summed up). For more than one variable, the Fokker-Planck equation becomes (Risken, 1984)

$$\frac{\partial p}{\partial t} = -\frac{\partial}{\partial x_i}\left(f_i(t,x_i)p\right) + \frac{1}{2}\frac{\partial^2}{\partial x_i \partial x_j}\left(g_{ij}^2(t,x_i)p\right). \tag{3.15}$$

Assuming an isotropic homogeneous diffusion process, the diffusion matrix reduces to:

$$\mathbf{D} = \begin{pmatrix} D & 0 \\ 0 & D \end{pmatrix}. \tag{3.16}$$

The drift function and noise function can again be determined from the Fokker-Planck equation (3.15) that corresponds to the advection-diffusion process:

$$f_i(t,x_i) = u_i + \frac{\partial D}{\partial x_i}, \tag{3.17}$$

$$g_i(t,x_i) = \sqrt{2D}. \tag{3.18}$$

The underlying sde that describes the advection-diffusion process is then equal to:

$$dX_{i,t} = \left(u_i + \frac{\partial D}{\partial x_i}\right)dt + \sqrt{2D}dW_{i,t}, \tag{3.19}$$

which is equal to the stochastic differential equation for the Fickian process (3.13) with an additional term $u_i dt$ for the advection. For the numerical implementation, we separated the sde equation into an advection part and a diffusion part. For the advection part, the equation for the particle displacement then collapses to a deterministic differential equation. For the diffusion part, we took the sde for the Fickian process (3.13). By splitting the equation, we were able to develop a very accurate numerical scheme for the advection part, which is especially needed for the LES model, where advection is the dominant term.

3.3 NUMERICAL IMPLEMENTATION ADVECTION

For the advection induced displacement of a particle, the following equation needs to be solved:

$$\frac{d\mathbf{x}}{dt} = \mathbf{u}, \tag{3.20}$$

where $\mathbf{u}$ represents the velocity of a particle, which depends on the position $\mathbf{x}$ of the particle. The vectors $\mathbf{x}$ and $\mathbf{u}$ contain the following entries:

$$\mathbf{x} = \begin{pmatrix} x \\ y \\ z \end{pmatrix}, \quad \mathbf{u} = \begin{pmatrix} u \\ v \\ w \end{pmatrix}. \tag{3.21}$$

It is assumed that the velocity varies linearly within a computational cell, so that the velocity can be written as:

$$\mathbf{u} = A\mathbf{x} + B, \tag{3.22}$$

where A represents a 3×3 matrix and B a 3×1 vector. These matrices are determined by substituting the positions and the velocities at the boundaries of the computational cell into the equation above. Three numerical schemes are discussed that solve eq. 3.20.

3.3.1 Euler scheme

The most straightforward numerical scheme is the Euler scheme, which uses the velocity at the previous time step to calculate the new position. So, this new position at time level $n+1$ is calculated by:

$$Y_i^{n+1} = Y_i^n + u_i^n \Delta t, \tag{3.23}$$

where Y_i represents the particle position for dimension i.

3.3.2 Runge-Kutta scheme

The Runge-Kutta scheme updates the velocity of the particle during a time step. The new position at time level $n+1$ is calculated by:

$$\begin{aligned}
&k_{1,i} = u_i^n \Delta t, && Y_i^{(1)} = Y_i^n + \tfrac{1}{2} k_{1,i} \Delta t, \\
&k_{2,i} = u_i^{(1)} \Delta t, && Y_i^{(2)} = Y_i^n + \tfrac{1}{2} k_{2,i} \Delta t, \\
&k_{3,i} = u_i^{(2)} \Delta t, && Y_i^{(3)} = Y_i^n + k_{3,i} \Delta t, \\
&k_{4,i} = u_i^{(3)} \Delta t, && \\
&Y_i^{n+1} = Y_i^n + \tfrac{1}{6} \left(k_{1,i} + 2k_{2,i} + 2k_{3,i} + k_{4,i} \right) \Delta t, &&
\end{aligned} \tag{3.24}$$

where $u_i^{(j)}$ represents the velocity at position $Y_i^{(j)}$.

3.3.3 Semi-analytical scheme

Also, a scheme based upon the analytic integration of the discretised velocity field was used (Dunsbergen, 1994; Ham et al., 2006). This method solves the streamline in a computational cell exactly. As a result, the streamlines are closed and particles can never cross closed boundaries. The expression for **u** is substituted into the original differential eq. 3.20, which results in a first-order system of differential equations:

$$\frac{d\mathbf{x}}{dt} = A\mathbf{x} + B, \tag{3.25}$$

The general solution for this system yields:

$$x_i = \sum_{j=1}^{3} c_j S_{ij} \exp(\lambda_j t), \tag{3.26}$$

where λ_1, λ_2 and λ_3 represents the eigenvalues with corresponding eigenvectors $S = [S_1, S_2, S_3]$ of matrix A. The values of c_j are determined from the initial condition:

$$\mathbf{x}(\mathbf{0}) = \mathbf{x_0}. \tag{3.27}$$

The particular solution is given by:

$$x_i^* = \sum_{j=1}^{3} S_{ij} \exp(\lambda_j t) \int_0^t \exp(-\lambda_j t) d_j(t) dt, \tag{3.28}$$

where the vector D is given by:

$$D = \begin{pmatrix} d_1 \\ d_2 \\ d_3 \end{pmatrix} = S^{-1}B. \tag{3.29}$$

Since d_j is independent of time, the particular solution can be written as:

$$x_i^* = \sum_{j=1}^{3} -\frac{d_j S_{ij}}{\lambda_j} \left(1 - \exp(\lambda_j t)\right), \tag{3.30}$$

which is zero at $t = 0$. The solution follows from the superposition of the general and particular solution:

$$x_i = \sum_{j=1}^{3} c_j S_{ij} \exp(\lambda_j t) + \sum_{j=1}^{3} -\frac{d_j S_{ij}}{\lambda_j} \left(1 - \exp(\lambda_j t)\right). \tag{3.31}$$

The exact numerical implementation depends on the mesh type used. We focused on the implementation for a Cartesian mesh and a tetrahedral mesh.

Cartesian mesh

The LES uses a Cartesian staggered mesh, which has clearly its advantages for the numerical implementation. The velocity in direction i only depends on the position x_i, so that the matrices A and S reduce to a diagonal matrix. The matrices A and B now contain 6 unknowns, which are solved by the 6 velocities at the boundaries of the cell. The solution for the particle position (3.31) then reduces to:

$$x_i = \left(\frac{d_i S_{ii}}{\lambda_i} + c_i S_{ii}\right)\exp(\lambda_i t) - \frac{d_i S_{ii}}{\lambda_i}. \tag{3.32}$$

Subsequently, the time is calculated when a particle would cross the cell boundary positioned at x_b, which can be solved easily from eq. 3.32:

$$\Delta t_i = \frac{1}{\lambda_i}\ln\left(\frac{\frac{x_b}{\lambda_i S_{ii}} + d_i}{d_i + c_i\lambda_i}\right). \tag{3.33}$$

These time steps are calculated for all the six boundaries of a computational cell. The smallest positive time step is chosen, which represents the time to cross this particular cell. The new position of the particle is computed by substituting this time step in eq. 3.32. This procedure is repeated for the adjacent cell where the particle has moved to.

Tetrahedral mesh

The RANS model with a standard k-ε model uses a tetrahedral mesh. In contrast with the Cartesian mesh, there is no direct expression for the time step at which a particle crosses the boundary of a cell. The time step needs to be solved by iteration, which increases the computational time considerably. The following calculation steps are needed:

- Transform the coordinate system in such a way that the four nodes of the tetrahedron are given by the coordinates (0,0,0), (1,0,0), (0,1,0) and (0,0,1). For each computational cell, a transformation matrix is calculated, which is used to transform the coordinates of the particle inside this particular cell.
- Determine the particle velocity equation for the computational cell (eq. 3.22). The matrices A and B are solved by substituting the 12 velocities (3 at each node of the computational cell) into eq. 3.22. Furthermore, the eigenvalues and eigenvectors of A are determined to evaluate the solution of the particle displacement (eq. 3.31).
- Calculate the new particle position. The time steps corresponding to the crossing of each face of the tetrahedron are calculated by iteration. The smallest positive time step is chosen to calculate the new particle position using eq. 3.31. The new positions are transformed back to the original coordinate system.

This procedure is repeated for the adjacent cell where the particle has moved to.

3.3.4 Time step requirements

The CFL (Courant, Friedrichs and Lewy) number can be used as a measure to define the time step (Courant et al., 1928). For the particle displacement, the CFL number is given by:

$$\mathrm{CFL} = \frac{u_p \Delta t_p}{\Delta x}, \tag{3.34}$$

where u_p represents the velocity of the particle (m/s), Δt_p the time step (s) and Δx the mesh size (m). This number is often set to a maximum value of 1, which means that a particle has a maximum displacement equal to the mesh size. For the semi-analytical method, this requirement is automatically fulfilled, since the particle displacement is updated when a particle crosses a computational cell. For the other methods, particle displacements may be very inaccurate if the CFL number exceeds 1. Moreover, this CFL requirement may be inefficient in the case of large variations in mesh sizes, because the small steps are not necessarily needed for the larger cells.

3.3.5 Test case: rotating flow

A 2D rotating flow was considered as a test case for the advection scheme. The velocity field was given by:

$$\mathbf{u} = \begin{pmatrix} -y \\ x \end{pmatrix}. \tag{3.35}$$

The streamlines should make a closed circle; the particles must return at a the same position after completing each cycle. In Figure 3.1, the particle paths for two particles, particle A with initial coordinates (−0.25,0) and particle B with coordinates (0.75,0) are shown for the three advection schemes. The Cartesian staggered grid was used. The maximum CFL number for the particles was set to 1, which occurred at the particle with the highest velocity (particle B). In other words, this particle crosses one computational cell at each time step. Since the semi-analytical scheme recalculated the particle path each time the particle crosses a cell boundary, a comparison with the other methods at CFL numbers higher than 1 would be unfair. The Euler scheme showed a very inaccurate solution, the particles had moved out of the domain. The Runge-Kutta scheme also resulted in inaccurate particle paths, the particles moved a little to the middle of the domain. Only the semi-analytical advection scheme was able to get an accurate solution of the particle path.

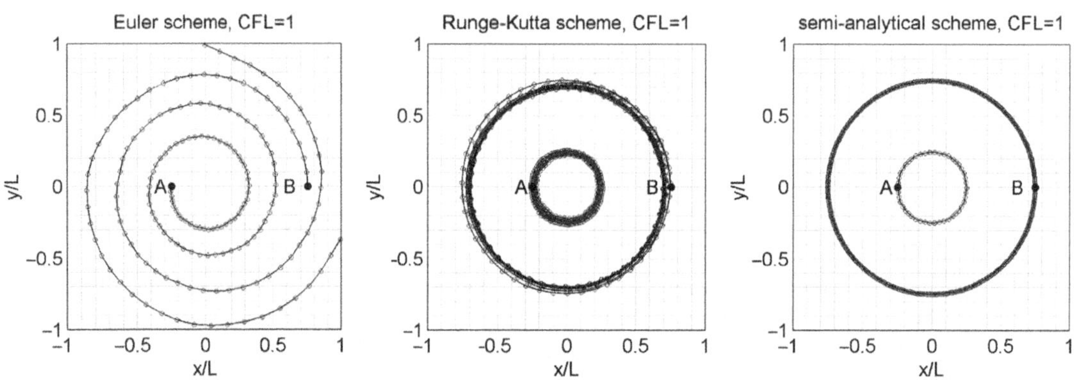

Figure 3.1 Comparison of different advection schemes for a rotating flow. The initial positions of particles A and B are shown by the black dot.

3.4 NUMERICAL IMPLEMENTATION DIFFUSION

Different schemes that approximate the Itô stochastic differential equation for the Fickian diffusion process (3.13) are discussed in this section. The schemes were derived from a stochastic Taylor expansion of the underlying sde (Stijnen, 2002; Charles, 2007). Adding up more terms of the Taylor expansion leads to more accurate schemes.

Explicit Euler scheme

The Euler scheme is relatively straightforward and includes only first-order terms. The new position is given by:

$$Y_{n+1} = Y_n + f(t_n, Y_n)\Delta t + g(t_n, Y_n)\Delta W_n. \tag{3.36}$$

For the Fickian diffusion process, the Euler scheme becomes:

$$Y_{n+1} = Y_n + \frac{dD}{dx}\Delta t + \sqrt{2D}\Delta W_n. \tag{3.37}$$

Milstein scheme

A higher order scheme is the Milstein scheme, where the second-order terms of the stochastic Taylor expansion are included (Kloeden and Platen, 1992). The new position of the particle then yields:

$$Y_{n+1} = Y_n + f(t_n, Y_n)\Delta t + g(t_n, Y_n)\Delta W_n + \frac{1}{2}g(t_n, Y_n)\frac{\partial g}{\partial x}\left((\Delta W_n)^2 - \Delta t\right). \tag{3.38}$$

For the Fickian diffusion process, the Milstein scheme becomes:

$$Y_{n+1} = Y_n + \frac{dD}{dx}\Delta t + \sqrt{2D}\Delta W_n + \frac{1}{2}\frac{dD}{dx}\left((\Delta W_n)^2 - \Delta t\right). \tag{3.39}$$

3.4.1 Test case: wall treatment

A challenge for diffusion schemes is to prevent particles from unrealistically crossing a closed wall. The two numerical schemes were therefore examined for a simple case with a non-uniform diffusion field. A two-dimensional square domain (40 × 40 elements) was chosen. In Figure 3.2 on the left, the noise term (g) is displayed. The diffusion was set to zero on the outer ring of the domain, so that $x = \pm L/2$ and $y = \pm L/2$ acted as virtual walls. Particles were not allowed to enter the outer region bounded by these virtual walls. In Figure 3.2 on the right, the drift term (f) is shown, which prevents the particles from accumulating at regions with low diffusion. Two different types of white noise were investigated: Gaussian white noise and uniform white noise, both processes had a random increment with mean zero and variance one. A total number of 1000 particles were released at the centre of the domain. The particles positions after 1000 time steps of $\Delta t = 0.1$ s are shown in Figure 3.3. The particles that had moved through the virtual walls are highlighted by a red circle. For the Euler scheme, almost all the particles had crossed the walls. The results were much better for the Milstein scheme. Using the Gaussian white noise a few particles had crossed the wall due to large increments in the Gaussian process, whereas using the uniform white noise no particles had crossed the wall. In contrast with the Gaussian white noise, for the uniform noise process the random increments are bounded (between $-\sqrt{3}$ and $\sqrt{3}$), so that particles were less likely to cross the wall. Of course, by further raising the time step or diffusion coefficient, particles cross the wall for all the methods, but it still holds that for the uniform white noise, particles are least likely to cross the wall. Using this method with not too large time steps (CFL not larger than 1), no special treatment, such as bouncing, is necessary at the wall. Therefore, the uniform white noise in combination with the Milstein scheme was used. Although a normal distribution would come closer to reality, since turbulent fluctuations are best represented by a normal distribution, no differences between a uniform or Gaussian noise were found in the case of a UV dose distribution prediction for the flow around a single cross-flow UV system (Chapter 5).

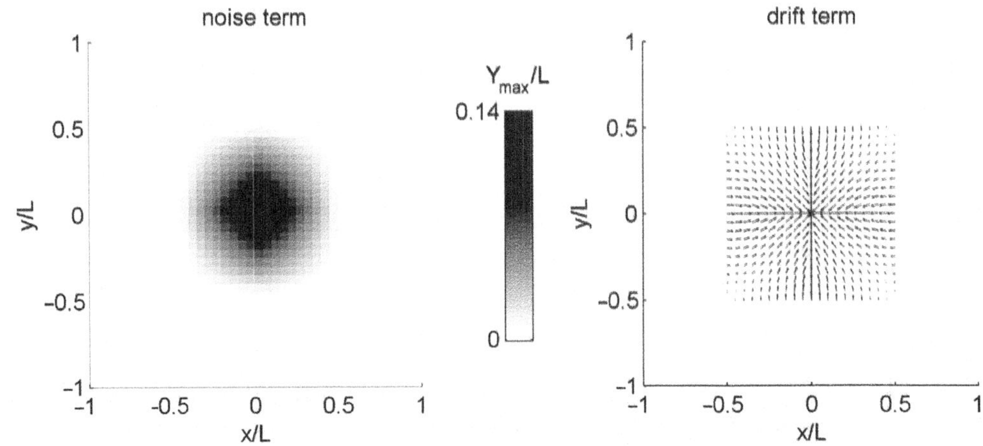

Figure 3.2 Left panel: Non-uniform diffusion coefficient ($Y_{max} = \sqrt{2D_{max}\Delta t}$). In the outer region, the diffusion was set to zero to simulate a wall. Right panel: Drift term corresponding to the diffusion field.

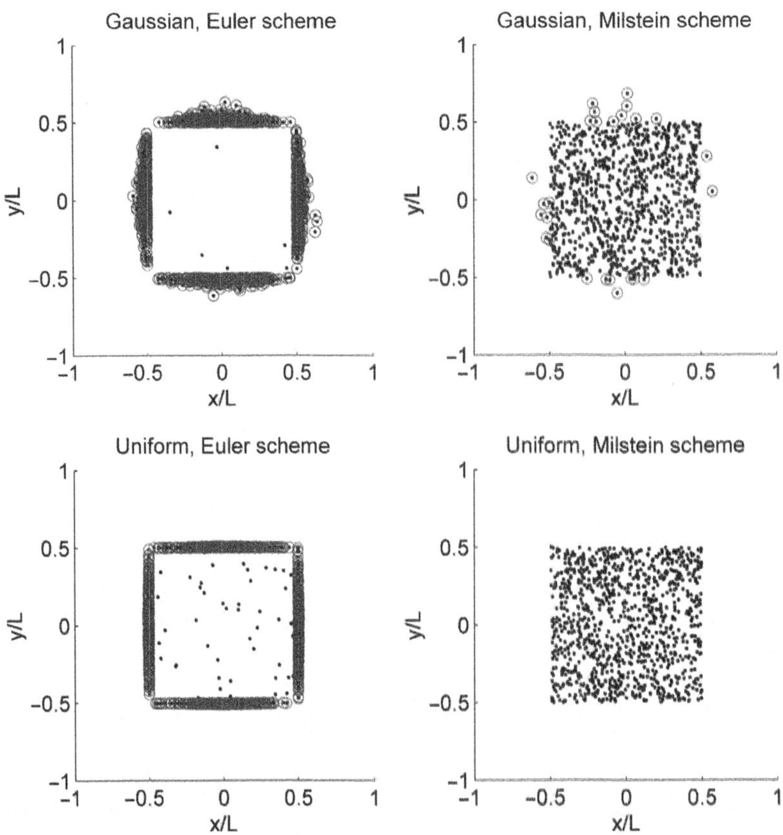

Figure 3.3 Particle distributions (in 2D) for different numerical schemes. The particles that had moved through the virtual walls are highlighted by a red circle.

3.4.2 Test case: diffusion properties

Two test cases were considered to evaluate the diffusion properties of the above mentioned Milstein scheme. A two-dimensional square domain was chosen (40 × 40 elements). For the first case, the diffusion coefficient was constant in x-direction and varied in y-direction (Figure 3.4). The maximum displacement was of the same order of magnitude as the element size. At the walls, the diffusion coefficients were set to zero, so no particles were allowed to cross the walls. Particles were initially uniformly distributed over the domain. The distribution must remain uniform, because there was no concentration gradient. With an inaccurate particle method the particles will accumulate in the regions with lower diffusion and cross the walls. But in the case of the Milstein scheme as implemented in eq. 3.39, the particle distribution should remain uniform. In Figure 3.4, the particle density in x- and y-direction is shown at different time steps. The particle distributions remained uniform, confirming that this method prevents particles from accumulating at regions with low diffusion and prevents particles from crossing the wall.

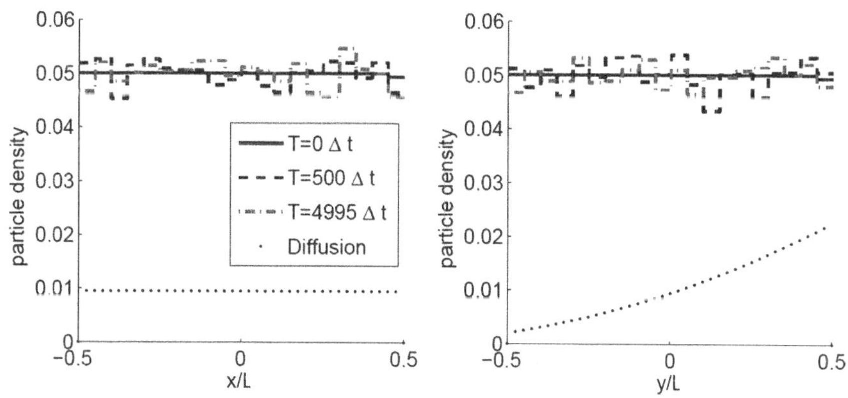

Figure 3.4 Particle distributions (in 2D) for a non-uniform diffusion coefficient in y-direction ($Y_{max} = 0.026L$, $N_p = 10000$).

For the second test case, the same domain was chosen, but the diffusion coefficient was set constant over the domain to evaluate the diffusion properties. After an initial release of particles in the middle of the domain, the distribution of particles must spread out according to a Gaussian distribution, given by:

$$p(x, y, t) = \frac{M}{\sqrt{4\pi Dt}} \exp\left(-\frac{x^2}{4Dt} - \frac{y^2}{4Dt}\right), \tag{3.40}$$

where M represents the initial mass. The analytical solution as well as the particle distribution are shown in Figure 3.5. Both distributions coincide, indicating that the diffusion properties were properly addressed by the numerical scheme.

3.5 TEST CASE: CHANNEL FLOW

The combination of an advection and diffusion displacement was tested for a simple channel flow, given by Elder (1959). Elder assumed a logarithmic (horizontal) velocity profile over the height ($u(y)$) equal to:

$$u(y) = u_m + \frac{u_*}{\kappa}\left(1 + \ln\left(\frac{y}{h}\right)\right), \tag{3.41}$$

and a parabolic diffusion profile $D(y)$:

$$D(y) = \kappa u_* y \left(1 - \frac{y}{h}\right), \tag{3.42}$$

where u_m is the mean velocity (m/s) and u_* is equal to the shear velocity ($u_* = \sqrt{c_f} u_m$, [m/s], friction coefficient $c_f = 3 \cdot 10^3$), h represents the water depth and κ represents the Von Karman constant ($\kappa = 0.41$). Elder used the theory of Taylor (1954) to derive a longitudinal dispersion coefficient (D_L, [m^2/s]):

$$D_L = 5.86 u_* h. \tag{3.43}$$

The spreading in longitudinal direction (averaged over the depth) is reproduced by a Gaussian distribution. The particle distribution over the depth is expected to become uniform after a peak release of particles over the middle of the height.

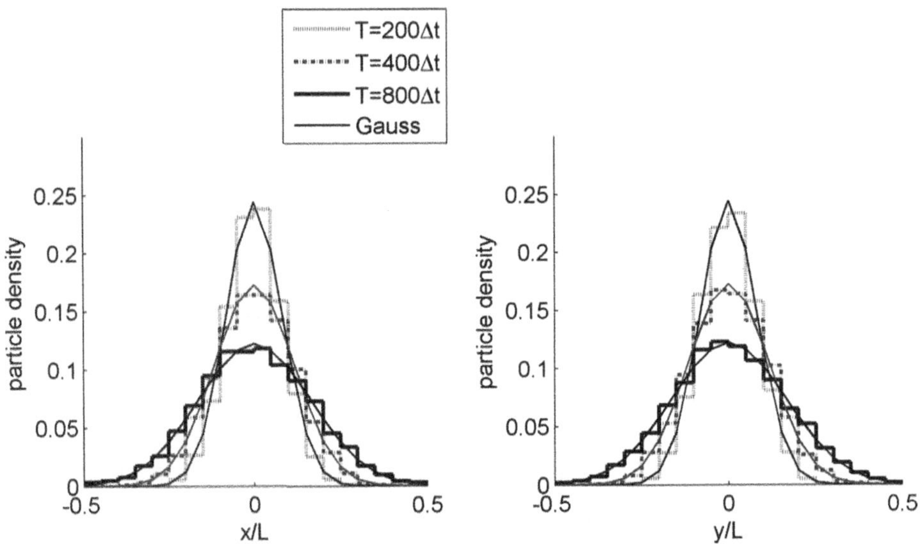

Figure 3.5 Particle distributions (in 2D) for a uniform diffusion coefficient ($Y_{max} = 0.01L$, $N_p = 10\,000$).

Particle displacements were simulated for the given velocity and diffusivity profile. A pulse injection of 10 000 particles at $x = 0$ and $y = 0$ (in the middle of the channel) was used. The particle distributions at different time steps are presented in Figure 3.6. The longitudinal particle positions were corrected with the mean velocity, so that a distribution was formed around the point $x - u_m t = 0$. After a while, the particles were uniformly distributed over the depth and approximated to a Gaussian distribution in longitudinal direction. No particles had crossed the wall. The dispersion coefficient converged to the theoretical value of Elder. The test case showed that the particle tracking technique for a combined advection and diffusion problem leads to accurate results.

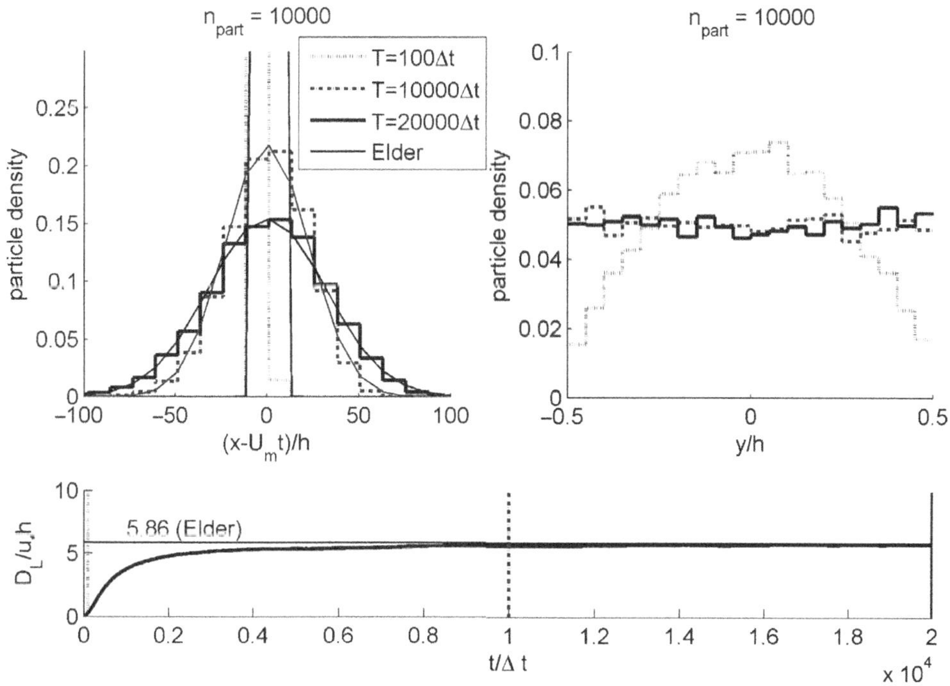

Figure 3.6 Particle distribution in x-direction and y-direction for a channel flow of Elder (1959), $h = 0.5\,\text{m}$, $\Delta t = 0.2\,\text{s}$.

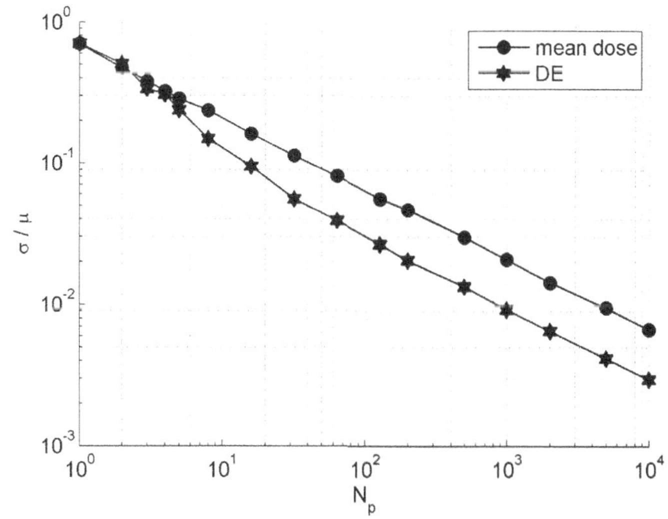

Figure 3.7 Number of particles versus the accuracy of the calculation. The latter is represented by the standard deviation (σ) divided by the mean (μ), which was determined for a large number of simulations with that particular number of particles (N_p).

3.6 NUMBER OF PARTICLES

The number of particles needs to be as small as possible to reduce computational time, but still sufficient to obtain accurate solutions. The number of particles required to come to an accurate prediction of the disinfection performance was therefore investigated. A number of particles N_p was drawn from a shifted exponential distribution (eq. 2.32), $\lambda = 0.01$ and $D_{\text{min}} = 50\ \text{mJ/cm}^2$, to simulate the dose distribution. This distribution, characterised by a D_{10}/D_{mean} of 0.41, represents one of the wider dose distributions occurring in reactors (Chapter 6). This provides a worst-case scenario for the required number of particles, because a wider distribution causes a larger variation between the doses of particles. The mean dose and disinfection were calculated from the simulated dose distribution. This was repeated for many times (1000 or larger), so that an accurate mean value and standard deviation was determined for the mean contact time and disinfection. The variation in computed mean dose or disinfection for that particular number of particles is then represented by the standard deviation divided by the mean. In Figure 3.7, this factor is plotted for different numbers of particles. This factor became smaller as the number of particles increases, resulting in a more accurate solution. For a number of 5000 particles, the standard deviation was only 0.5% of the mean for disinfection (*DE*), which was regarded as accurately enough to predict the disinfection.

For the residence time distribution, it was also shown that a number of 5000 particles resulted in accurate results with respect to an Eulerian approach (§4.3.2).

Chapter 4

Ozone systems*

Ozone systems are evaluated in this chapter using the modelling tools described in Chapter 2 and 3. The validation of a full-scale installation by residence time distribution measurements is discussed, and hydraulic optimisations of ozone contactors are evaluated. An assessment is made of different disinfection calculation methods to come to an accurate disinfection prediction.

4.1 INTRODUCTION

Worldwide, over 3000 ozone contactors are used in drinking water treatment plants, primarily for disinfection purposes. Other benefits of ozone treatment besides disinfection are: oxidation of reduced metals, reduction in taste and odour, decolouration of the water, enhanced coagulation and reduction in trihalomethane formation (von Gunten, 2003b). Ozone also inactivates chlorine resistant micro-organisms like *Gardia* and *Cryptosporidium* (von Gunten, 2003a). The disinfection efficiency of the ozone systems depends on the ozone dose (or CT value) distribution. Due to suboptimal hydraulics, different parts of the water receive a different ozone dose, which results in such a distribution. The low ozone doses are very critical for the disinfection due to the high extent of removal required (for example, 3 log or 99.9%). Therefore, hydraulic optimisations that increase the lowest ozone doses are very beneficial for the disinfection. A disadvantage of the use of ozone is the formation of disinfection by-products, especially the production of bromate in bromide containing water is a point of concern (von Gunten, 2003a). Although discussions are ongoing on the presumed toxicity of bromate (Cotruvo et al., 2009), bromate control is still an important issue. By Dutch legislation a maximum bromate concentration of 5 μg/L is allowed (90th percentile), or 10 μg/L if ozone is used as a primary disinfection barrier (VROM, 2001). Bromate issues as well as economical considerations require optimisations of ozone contactors that reduce the ozone dosage, whereas the microbial safety is guaranteed or even improved. It was shown by van der Helm et al. (2007)

---

* Parts of the research described in this chapter are published in: Wols, B.A., Hofman, J.A.M.H, Uijttewaal, W.S.J., Rietveld, L.C., Stelling, G.S. and van Dijk, J.C. (2008). Residence time distributions in ozone contactors. *Ozone: science and engineering*, 30(1):49–57, http://dx.doi.org/10.1080/01919510701759538
Wols, B.A., Hofman, J.A.M.H, Uijttewaal, W.S.J., Rietveld, L.C. and van Dijk, J.C. (2010). Evaluation of different disinfection calculation methods using CFD. *Environmental modelling and software*, 25(4):573–582, http://dx.doi.org/10.1016/j.envsoft.2009.09.007
The data can be accessed through http://dx.doi.org/10.4121/uuid:c1ac7344-1419-4398-ba13-c757551c303f

that the bromate formation has a linear relation with the average CT value. An accurate prediction of the disinfection is therefore of major interest, so that the required ozone dosage can be determined precisely. As a result, no excess of ozone is added to the water, which reduces the formation of disinfection-by-products as well as the energy consumption. Moreover, an optimisation of the hydraulics may narrow the ozone dose distribution, so that the same disinfection can be reached at a lower ozon dosage. Such an optimisation reduces the production of bromate, whereas the microbial safety remains ensured. The ozone contactor consists usually of a bubble column, where the ozone gas is injected and dissolved in the water, and a number of contact chambers to ensure enough reaction time of the dissolved ozone with the contaminants (Figure 4.1).

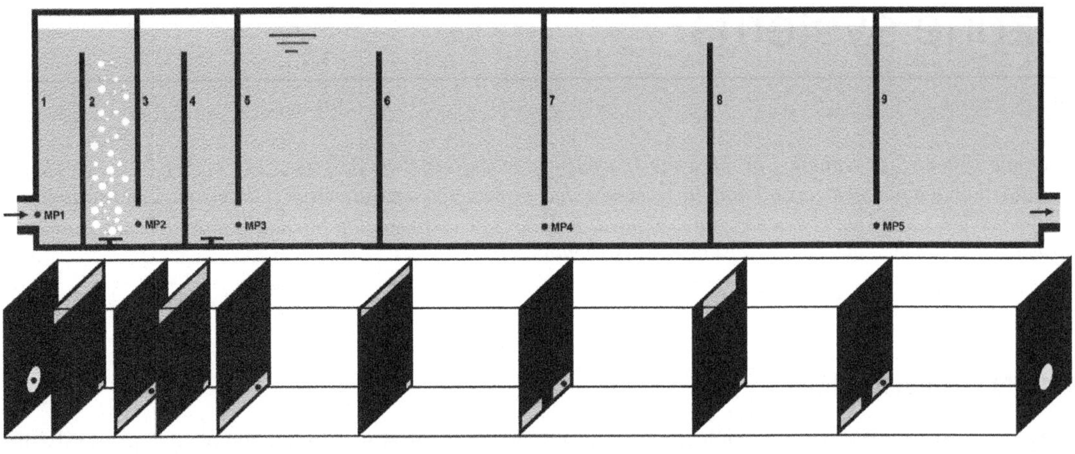

Figure 4.1 Geometry of the ozone contactor at the Leiduin WTP, denoted as reference contactor.

The geometry of the contactor causes a complex flow pattern that requires sophisticated modelling. Therefore, CFD modelling is used to predict the disinfection in ozone contactors. Several authors used CFD as a tool to optimise ozone contactors (Cockx et al., 2001; Huang et al., 2004; Li et al., 2006; Zhang et al., 2007). They showed that improved hydraulics and/or microbial inactivation are obtained by changing the contactor geometry. However, the disinfection was calculated in different ways. Cockx et al. (1999) and Zhang et al. (2007) used the CT concept and calculated the CT value by solving the scalar transport equations (concerning advection, diffusion and a source term), from which they determined the inactivation from CT tables (USEPA, 1991, 2006b). Huang et al. (2004) calculated the microbial inactivation directly by solving the scalar transport equations for a concentration of micro-organisms. Li et al. (2006) used a particle tracking technique to determine the residence time distribution, but no disinfection calculation was made here. These methods were explained in §2.6.

First of all, a residence time distribution experiment conducted at the Leiduin WTP of Waternet is discussed in §4.2, which acted as a validation for the CFD modelling. This validation as well as some modelling aspects to get to a proper modelling result are discussed in §4.3. Using the validated CFD model, modifications to the geometry of the ozone contactor that optimise the hydraulics were investigated (§4.4). Furthermore, the above mentioned calculation methods and some other methods described in Chapter 2 were evaluated (§4.5). CFD was used as a tool to assess the disinfection calculation methods, because it determined all the necessary quantities (residence time distributions, ozone concentrations, particle trajectories, CT values) needed for the disinfection calculation, so that a reliable comparison could be made. The most complex method, the particle tracking method is expected to result in the most accurate disinfection prediction, since it comes closest to the actual movement of individual micro-organisms. Additionally, a sensitivity analysis with respect to the kinetic constants (slow ozone decay and inactivation rate constant) was conducted (§4.6). Finally, issues regarding short-circuiting of water parcels in the contactor are discussed in §4.7.

4.2 EXPERIMENTS OF LEIDUIN OZONE CONTACTOR

4.2.1 Setup

Residence time distributions were measured at different positions in the full-scale ozone contactor at the water treatment plant Leiduin, from now on defined as reference contactor (Figure 4.1). Here, four parallel ozone contactors are operated, of which one was used for the experiment. Fluoride was used as a tracer, because low concentrations could be measured without adverse effects of the quality of the drinking water. The maximum fluoride concentration in one ozone contactor was 2 g/m^3. A step injection of fluoride was added to the feed pipe, while in the meantime at 5 locations underneath the baffles samples were taken (with intervals of 0.5, 1 or 2 minutes). Measurements took around 4 times the hydraulic residence time to ensure sufficient time to fill up the recirculation zones in the compartments. Measurements were conducted at flow rates of 1200, 2000 and 2500 m^3/h.

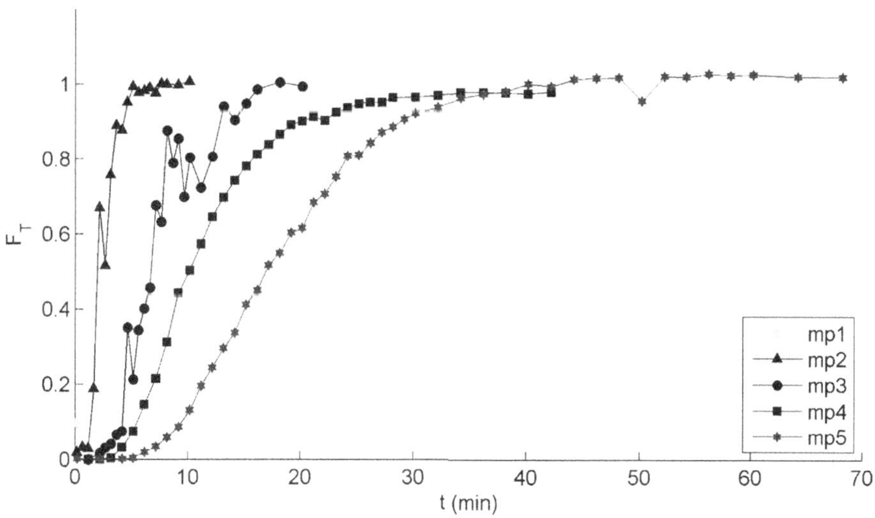

Figure 4.2 Measured cumulative residence time distributions in the reference ozone contactor at a flow rate of 2000 m^3/h.

Table 4.1 Residence times at several measuring points (MP) in the tracer experiment.

MP	Q (m^3/h)	T_h(min)	T_{10}/T_h	T_{50}/T_h	T_{90}/T_h	T_m/T_h	Pe
mp2	2001	2.3	0.37	0.86	1.71	0.99	4.5
mp3	2001	4.6	0.70	1.39	2.92	1.57	7.7
mp4	2001	10.7	0.48	0.92	1.85	1.04	6.9
mp5	2001	17.5	0.51	0.96	1.63	1.06	8.0

4.2.2 Residence time distributions

The normalised residence time did not show significant differences between the flow rates, so only the 2000 m^3/h flow rate was considered. The measured residence times are presented in Figure 4.2 and Table 4.1. The first measuring point (mp1) located at the entrance of the contactor showed a higher concentration than expected, indicating that the fluoride was not well mixed yet. The fluoride was, however, well mixed after the bubble column at measuring point 2, where the same maximum concentration was reached as in the following measuring points. Measuring point 3 showed large fluctuations in the fluoride concentration. Also, the mean residence time (T_m) was larger than what would be expected from the hydraulic residence time (T_h). A possible explanation is that a small hole in the upstream baffle (a safety measure to empty the contactor if necessary) generated a small recirculation zone, inside which the measuring point might be located. This could not be verified due to limited access to the contactor. However, the other measuring points did show a consistent behaviour – the concentration profiles were smooth and the mean residence times corresponded with the hydraulic residence time. Therefore, measuring points 4 and 5 were used for validation purposes. They showed a smooth increase towards the maximum concentration. The residence time distribution became wider with increasing length over the contactor. After each compartment, the combination of short-circuit flow and recirculation zone was responsible for a widening of the residence time distribution. The T_{10} was of the order of 50% of the hydraulic residence time (Table 4.1), indicating that there was a significant amount of short-circuiting.

4.3 CFD MODELLING OF VARIOUS OZONE CONTACTORS

4.3.1 Modelling conditions

The finite-element model Finlab (Labeur and Wells, 2007) was used for the CFD modelling. The CFD model gave predictions of flow fields, ozone concentration fields, residence time distributions, particle trajectories, CT values and inactivation levels. For the calculation of the flow fields, the inflow boundary was set as an inflow velocity equal to 0.144 m/s (which corresponds to a flow rate of 2000 m^3/h). The outflow boundary was modelled as a pressure in correspondence with the level of the free water surface ($z = 5.6\,m$). At the walls a velocity in accordance with a logarithmic velocity profile was prescribed (corresponding to a Nikuradse wall roughness of 10^{-4} m). The free-surface was modelled as a rigid lid (free-slip conditions). For the transport equations, the inflow was modelled as a concentration boundary and the outflow was modelled as a convective flux boundary. The walls and free-surface were regarded as zero flux boundaries. The ozone decay was modelled as a first-order process with a (slow) ozone decay coefficient of $k_s = 2.5 \cdot 10^{-3}\ s^{-1}$. For disinfection, we introduce the factor $k_\mu C_0$, the inactivation rate constant times the initial ozone concentration, which was set to 0.8 min^{-1}. The numerical constants used in the CFD model are given in Table 4.2. A time step of 0.1 s for the flow calculation was used, whereas the less complex scalar transport calculation sufficed with a time step of 0.25 s. For the particle tracking, a total number of 5000 particles was sufficient to obtain convergence (spreading in inactivation levels between particle sets was below 1% of the mean value). The time step was chosen such that a particle would travel a distance of around one element.

4.3.2 RTD: Eulerian versus Lagrangian

A passive tracer was added to the inflow boundary by means of a step-wise increase of the concentration (Eulerian approach). The concentration spread out in flow direction and the response to the step function gave the cumulative residence time distribution at several locations. For the Lagrangian simulation, the cumulative particle distribution after a pulse injection of particles is presented (a step injection requires too much particles and therefore too much computational effort). The particles were released at the inlet, initially at rest, uniformly distributed over the height. A number of 1000 particles gave some variations in the particle distribution (Figure 4.3), but convergence was already obtained for an injection of 5000 particles. The particle distribution appeared to approximate the Eulerian solution.

Table 4.2 Numerical constants used for the CFD modelling.

	Reference	No baffles	Add baffles	Turning vanes
Element type	Triangular	Triangular	Triangular	Triangular
Max. element size	0.266 m	0.226 m	0.359 m	0.271 m
Min. element size	0.017 m	0.030 m	0.027 m	0.019 m
Number of elements	19 030	20 998	28 928	31 444
Time step flow solver	0.1 s	0.1 s	0.1 s	0.1 s
Time step scalar transport	0.25 s	0.25 s	0.25 s	0.25 s
Number of particles	5000	5000	5000	5000
Time step particles	0.2 s	0.2 s	0.2 s	0.2 s

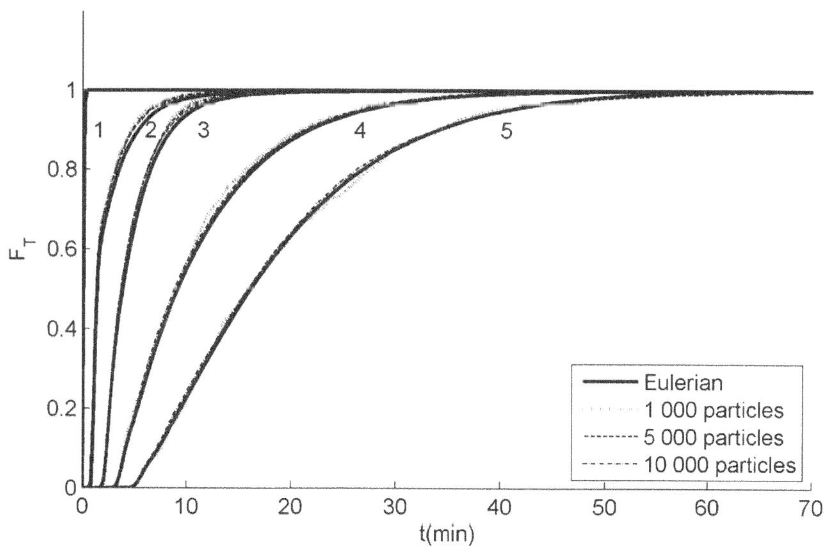

Figure 4.3 Cumulative residence time distributions calculated by the Lagrangian method for different numbers of particles and by the Eulerian method. The results for measuring points 1 to 5 are shown from left to right.

4.3.3 Validation of residence time distributions

A validated residence time distribution indicates that the CFD model properly assesses the hydraulics, which gives confidence in the capabilities of the CFD model to predict other properties, such as ozone concentrations, CT values etc., correctly. In Figure 4.4, the measured residence time distribution underneath the last baffle in the contactor as well as the computed values for a 2D and a 3D CFD model are plotted. The 2D model agreed reasonably well with the measurements, but not as good as the 3D model. The 3D flow effects enhanced the mixing over the cross-section, so that the RTD was reduced a little in comparison with the 2D model and eventually showed a better agreement with the measured data. To reduce computational costs, the 2D model was chosen to investigate geometric modifications to the contactor (which are invariant with respect to the third dimension) and for the assessment of different disinfection methods. All the disinfection methods (except for the CSTR method) used the same data from the CFD model, so that still a sound assessment of the disinfection methods could be made.

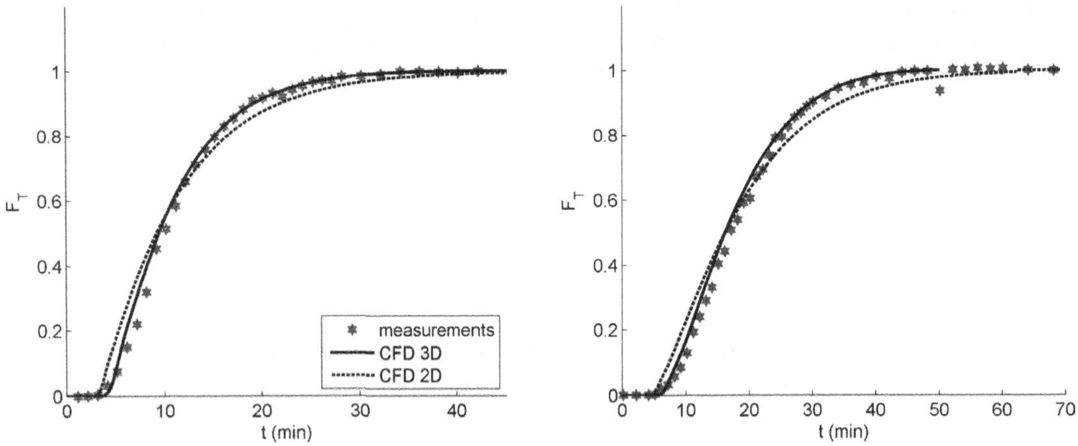

Figure 4.4 Cumulative residence time distribution, calculated by the CFD model, were validated by measurements. Left panel: mp4, right panel: mp5.

4.3.4 Mesh independency

The mesh needs to be sufficiently fine to obtain accurate results. Since we are interested in disinfection, we consider the predicted disinfection to investigate the mesh independency. For the contactor without baffles (§4.4), using at least 20 000 elements was enough to ensure mesh independency. Decreasing the mesh sizes resulted in similar log inactivation levels, calculated by the particle tracking method (Figure 4.5). The other contactors (§4.4) use similar mesh sizes or smaller.

4.4 CFD RESULTS OF HYDRAULIC OPTIMISATIONS

Four baffle contactor variants with similar outer dimensions were considered (length = 38 m, height = 5.2 m, width = 3.9 m), so that mean residence times were the same. The ozone contactor at the Leiduin WTP of Waternet (Amsterdam Water Supply) was taken as a reference. A flow rate of 2000 m^3/h per contactor was used for the CFD calculations. A configuration without baffles after the bubble column was considered to investigate the effect of the vertical baffles in the reference contactor. Geometries with additional baffles and turning vanes that reduce short-circuiting were also investigated.

Besides the contactor systems, a plug flow system was considered, because it represents an ideal hydraulic configuration. Comparing the contactors with suboptimal hydraulics to the ideal configuration showed the potential improvement in disinfection capacity. The plug flow was modelled as a flow between two plates (length = 200 m, height = 2 m). The ozone was injected 100 m downstream of the reactor inlet, so the velocity profile between the plates was expected to be fully developed. Mean residence times defined from the injection of ozone to the outlet of the reactor were in the same range as for the baffle contactors.

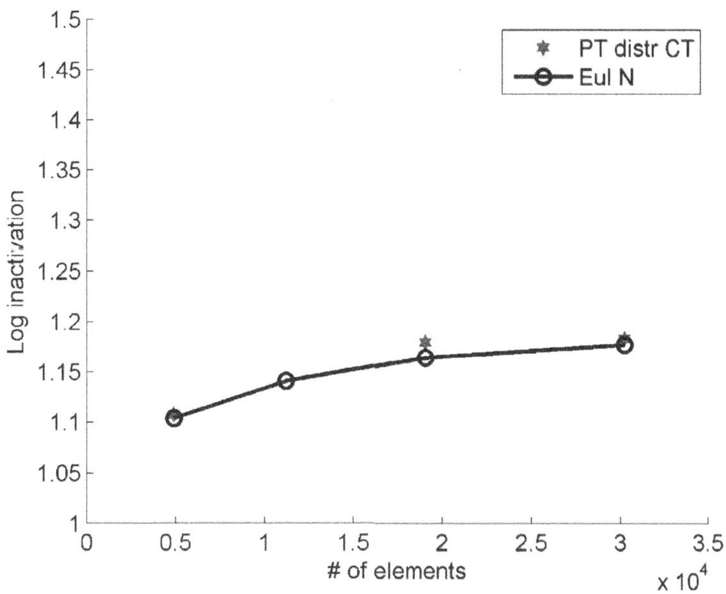

Figure 4.5 Log inactivation levels for the contactor without baffles as a function of the number of mesh elements ($k_S = 2.5 \cdot 10^{-3}\ s^{-1}$, $k_{\mu}C_0 = 0.8\ min^{-1}$).

Flow field and ozone concentration

In Figure 4.6 on the left, the flow fields are depicted for the contactor variants. For the reference contactor, the velocities increased above and below the baffles. This resulted in regions with high velocities that caused short-circuits. At the fifth baffle, velocities were even higher, because this baffle is higher than the other baffles. Flow separation occurred downstream of the baffles resulting in a large recirculation zone. In the case of no baffles a very large recirculation zone developed that occupied the complete contactor. The high velocities underneath the baffle were reduced by the shear layer that had spread out over the height. Additional baffles reduced the strength of the large recirculation zones as well as the strength of the short-circuit flows. The turning vanes had similar effects.

In Figure 4.6 on the right, the ozone concentrations for a decay coefficient of $2.5 \cdot 10^{-3}\ s^{-1}$ are depicted. By assuming a complete mixing between (dissolved) ozone and water after the bubble column, a uniform (dissolved) ozone distribution underneath the baffle upstream was set as an initial condition. Increasing the number of baffles and therefore reducing the size of the recirculation zones resulted in a more uniform distribution of the dissolved ozone in the contactor.

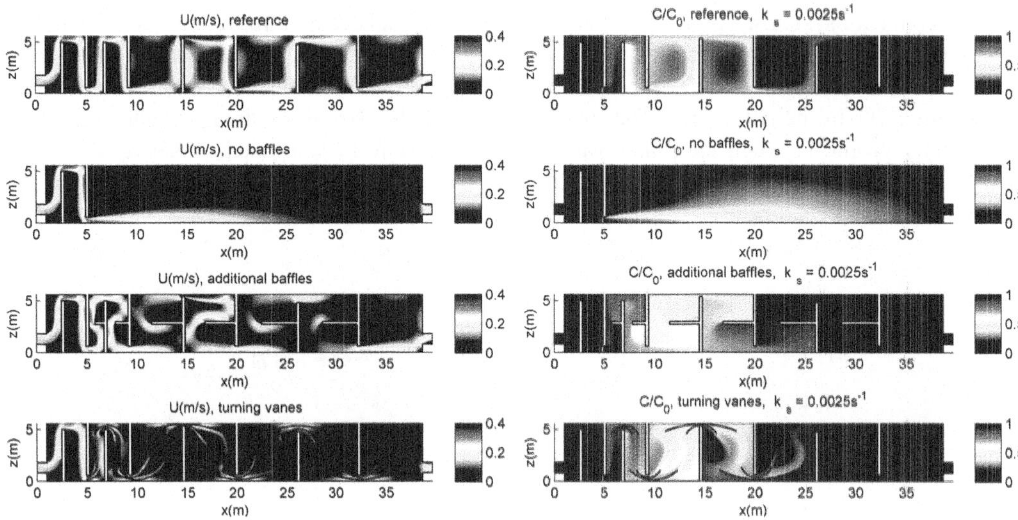

Figure 4.6 Flow fields (left panel) and ozone concentrations (right panel) for alternative contactor geometries.

Residence time distribution

The (cumulative) residence time distributions (RTD) at the effluent are plotted in Figure 4.7. The contactor without baffles showed the largest spreading in residence time. Especially the larger residence times increased with respect to the reference contactor, whereas the smallest residence times were in the same range. Although the flow fields between the reference contactor and the contactor without baffles were completely different, the influence of the vertical baffles on the residence time distribution was relatively small. So, the differences in flow field only had a small effect on an integrated property like the residence time distribution. The adjustments to the contactor, proposed as additional horizontal baffles or turning vanes, had a larger effect: the residence time distribution became narrower, the T_{10} was almost doubled. The variation in residence time was much smaller for the plug flow system, where the (cumulative) RTD approximated a step function.

CT values

The ozone exposure (CT value) was calculated for every particle released in the flow domain (by integrating the ozone concentration over the particle's trajectory). A number of 5000 particles were released instantaneously at the inlet. While the particles moved towards the outlet of the contactor, the ozone exposure of each particle accumulated until the end of the installation was reached. In Figure 4.8, the particle positions 5.57 minutes after a peak release of all the particles are depicted for the different geometries (CT value is indicated by colour). The CT value was scaled with the inflow concentration of dissolved ozone and the decay coefficient. At this instant, particles inside the reference contactor had just reached the outlet. These 'quick' particles had received low ozone dosages and, therefore, were critical for the disinfection performance. For the configuration without baffles, the 'quickest' particles had not yet reached the outlet, because the absence of baffles allowed for a large shear layer that caused a considerable deceleration of the main flow. However, a larger group of particles with slightly lower CT values than in case of the reference contactor reached the outlet. For the geometries with additional baffles or turning vanes, the 'quickest' particles only reached half of the contactor, whereas CT values were already higher.

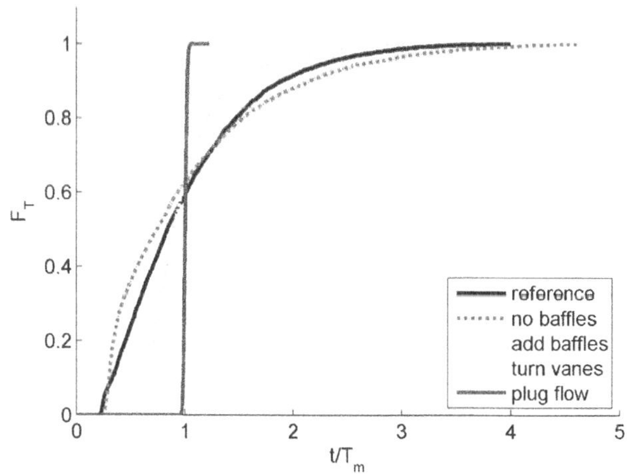

Figure 4.7 Cumulative residence time distributions for the different contactor geometries. Residence times are normalised with the mean residence time.

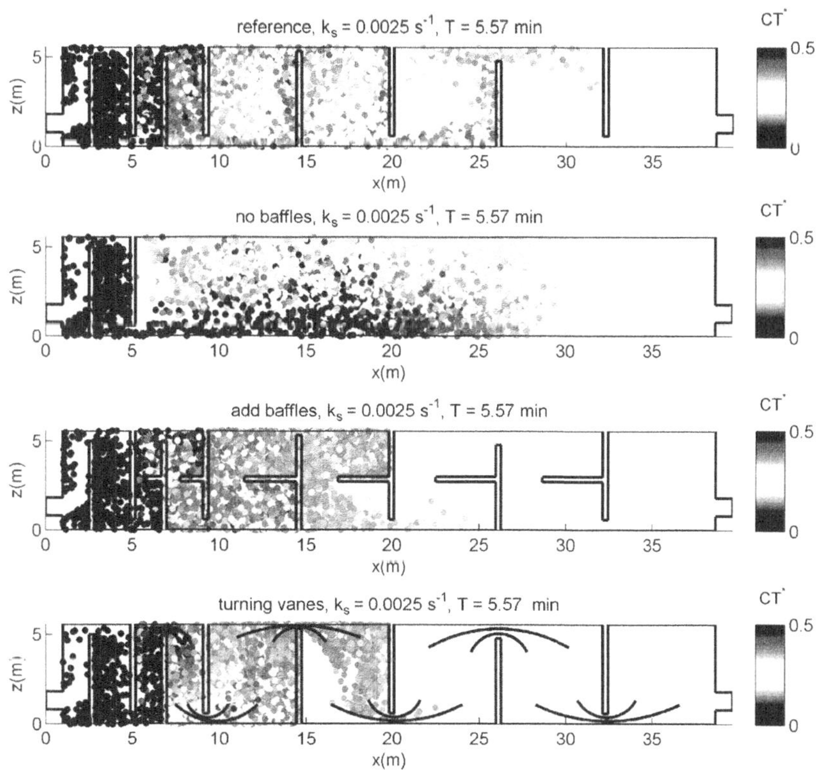

Figure 4.8 Particle positions 5.57 minutes after a peak release of particles at the inlet. The normalised CT value, $CT^* = CT \cdot k_S/C_0$, is indicated by the colour of the particle. A video can be found at: http://dx.doi.org/10.4121/uuid:f284a6f2-04a2-431e-9896-4a04bd61ac6b.

The CT value was recorded for every particle reaching the outlet, from which a distribution of CT values was constructed (Figure 4.9). The distribution of CT values provided a complete overview of the performance of these ozone contactors. For the geometry without baffles, a peak occurred at the smallest CT values, caused by the (large number of) particles that left the contactor quickly through short-circuiting. Very high CT values developed for particles that were trapped inside the large recirculation zone. For the reference configuration a narrower CT value distribution developed: the peak occurred at higher CT values, although the distribution was still asymmetrical and had a large tail at higher CT values. Improving the hydraulics by adding baffles or turning vanes narrowed the CT value distribution and shifted the peak towards a higher CT value. The distribution of CT values was obviously the narrowest for the plug flow configurations.

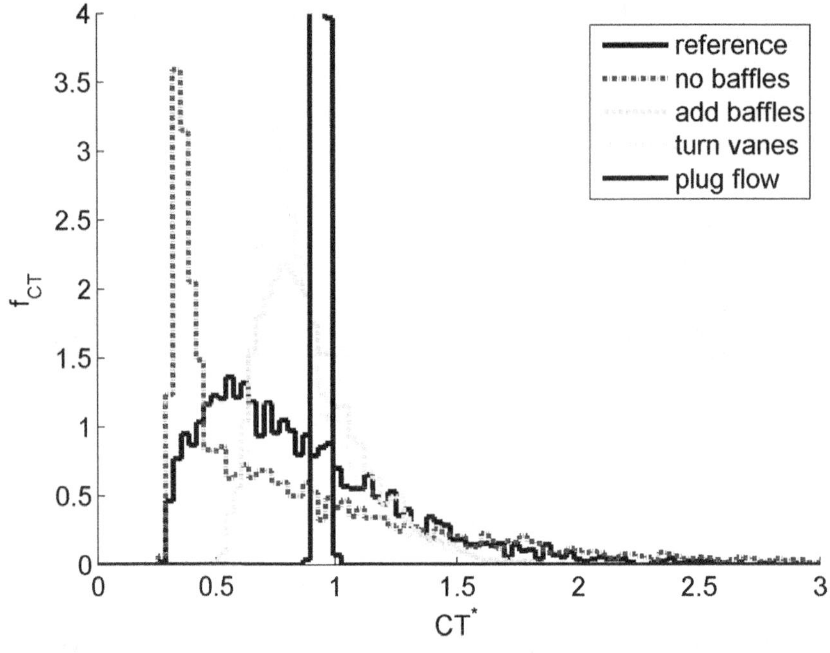

Figure 4.9 Distribution of normalised CT values obtained from the ozone exposures of the particles at the end of the contactor ($k_S = 2.5 \cdot 10^{-3} s^{-1}$).

The ozone exposure was also calculated by the Eulerian mean CT value method (§2.4.3). A spatial distribution of time averaged CT values over the contactor was obtained (Figure 4.10, left). The CT value at the outlet of the contactor was almost uniformly distributed over the height. Differences between the various geometries were small: the mean CT values at the outlet were almost equal. In Table 4.3, the mean CT values calculated by the particle method were compared to the mean CT value defined by the Eulerian method (averaged over the outlet boundary). These mean CT values were almost the same, confirming that these methods were consistent with each other. However, the fact that the mean CT values of the different geometries were almost the same, whereas the distribution of CT values largely differed (Figure 4.9), shows that a proper disinfection calculation is required that accounts for the distribution of CT values.

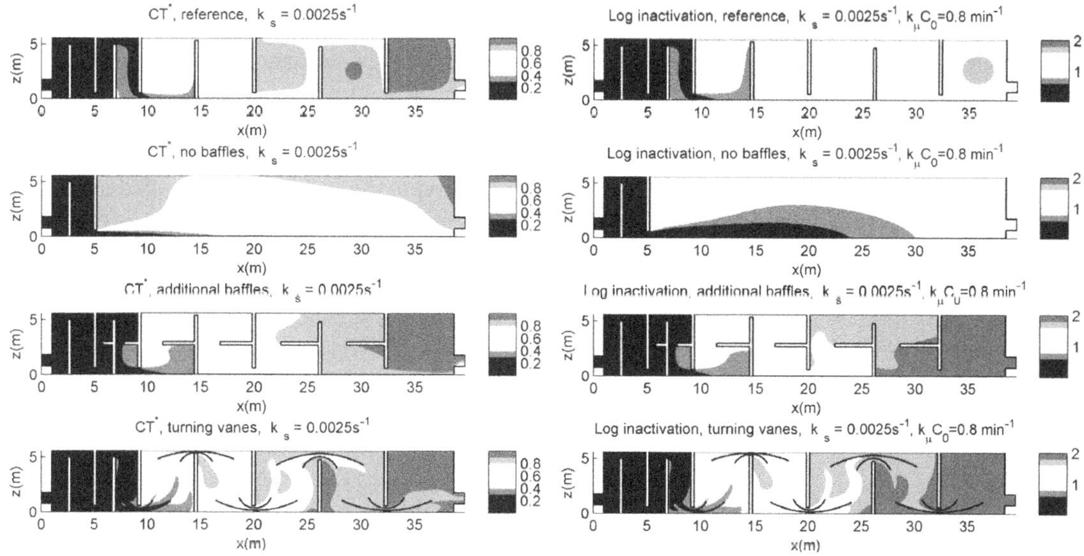

Figure 4.10 Mean CT values (left) and log inactivation levels (right) calculated by an Eulerian method.

Table 4.3 CT values (normalised) and log inactivations (*DE*) calculated by a Lagrangian and Eulerian method.

	Reference	No baffles	Add baffles	Turning vanes	Plug flow
CT^*_{mean} Lagrangian	0.83	0.85	0.90	0.88	0.94
CT^*_{mean} Eulerian	0.84	0.86	0.91	0.90	0.95
DE Lagrangian	1.43	1.18	1.88	1.86	2.18
DE Eulerian	1.44	1.16	1.89	1.90	2.17

$k_s = 2.5 \cdot 10^{-3}$ s^{-1}, $k_\mu C_0 = 0.8$ min^{-1}, $DE_{max} = 2.22$.

4.5 ASSESSMENT OF DISINFECTION MODELS

Differences between disinfection models for the reference contactor

Disinfection calculated by the particle tracking method (PT distr CT) was regarded as the most accurate prediction, because it comes closest to the actual movements of micro-organisms. The different disinfection models (§2.6) were compared to the particle tracking method for the reference ozone contactor. The CT_{10} method predicted lower log inactivation levels than the particle tracking method (Figure 4.11). For the CSTR method, the number of CSTRs was equal to the number of compartments. This method almost showed the same disinfection prediction as the particle tracking method. The segregated flow analysis (SFA) showed much higher disinfection levels, whereas the micro-mixing analysis (MMA) showed lower disinfection levels than the particle tracking method. The Eulerian mean CT value method (Eul mean CT) gave the highest disinfection levels. The log inactivation levels determined by the Eulerian direct inactivation method (Eul N), also shown in Figure 4.10 on the right, approached the disinfection levels calculated by the particle tracking method best. Conservative methods, such as CT_{10} method and MMA, are suitable for calculation, because the microbial safety is ensured. However, these methods overestimate the required ozone dosage, and, consequently, the production of disinfection by-products. The Eulerian mean CT value approach and SFA should be used with caution, because these methods may overestimate the disinfection performance. As indicated by Greene et al. (2006), the information given by the residence time distribution is insufficient, because it does not contain the actual mixing state. Extremes in mixing states are represented by the MMA and SFA. The MMA assumes a complete mixing between the water segments, which resulted in an underestimation of the disinfection capacity. The assumption made in the SFA that elements of water do not mix with each other may lead to the unrealistic situation that all flow segments will receive the same CT values (when residence times compared to the ozone decay are long enough). Because mixing between the flow segments results in a larger distribution of CT values than predicted by the SFA, the SFA will overestimate the disinfection capacity. The Eulerian direct inactivation method does provide a proper disinfection calculation, whereas the computational costs are lower than for the particle tracking method. However, the particle tracking method provides statistics of ozone exposures, which is useful information for the design of ozone systems. Moreover, once the particle trajectories are known, alternative kinetics can be assessed quickly.

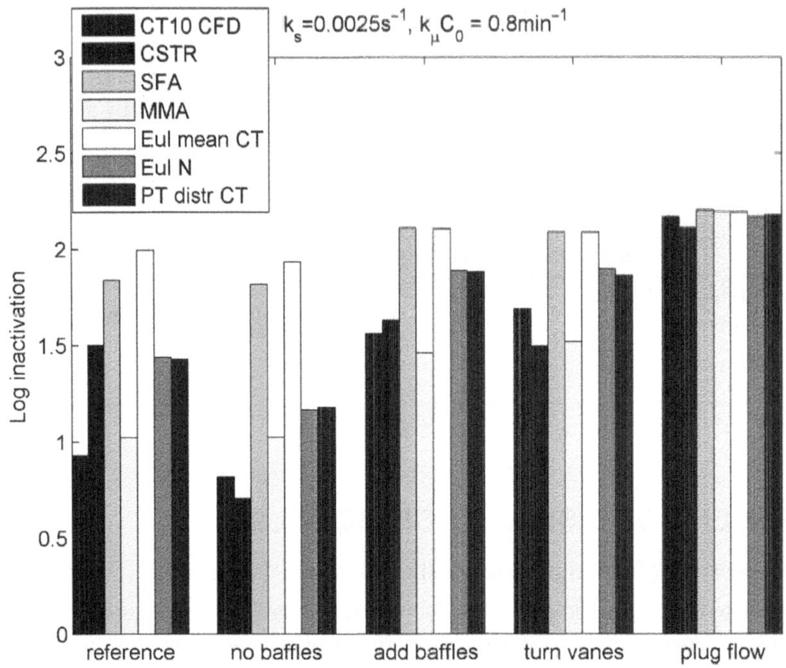

Figure 4.11 Log inactivation levels predicted by the various disinfection calculation methods for the different geometries.

Effect of hydraulic optimisations

An overview of the log inactivation levels determined from the different disinfection calculation methods is shown in Figure 4.11. The most accurate methods, particle tracking and Eulerian direct inactivation method, showed an increase in disinfection level by optimising the hydraulics. Hydraulic improvements such as additional baffles or turning vanes reduced the number of small CT values and therefore increased the disinfection capacity, whereas the mean CT value was not changed. The disinfection levels for the contactors with these geometric changes came close to the ones of the plug flow system, where the hydraulics were most optimal. A contactor without baffles showed lower inactivation levels, mainly because the sudden expansion from the small cross-section of the inflow to the large cross-section of the contactor resulted in a large recirculation zone.

Most of the disinfection models predicted higher disinfection levels when optimising the hydraulics. For the CSTR method, an increase in disinfection levels occurred, because the number of CSTRs was increased, which resulted in more efficient hydraulics. For the plug flow reactor a number of 100 CSTRs was chosen, which resulted in disinfection levels in accordance with the other disinfection calculation methods for such a system. However, for the other configurations, due to the limited hydrodynamics incorporated in the CSTR method, the CSTR method deviated more pronouncedly from the particle tracking method. For the segregated flow analysis, changing the geometry had little effect on the disinfection level. Also, the Eulerian mean CT value method (Eul mean CT) did not predict the improved disinfection by the hydraulic adjustments, because the mean CT values were equal. All the disinfection calculation methods showed the same results for the plug flow configuration, where the log reduction values were higher than for the other configurations due to the narrowness of the CT value distribution.

4.6 SENSITIVITY TO KINETIC PARAMETERS

An overview of the log inactivation levels, calculated by the particle tracking method, as a function of the decay coefficient is given in Figure 4.12. This range of decay coefficients was measured for the reference ozone contactor (van der Helm et al., 2007). The black dash-dot line represents the maximum log inactivation, which is an indicator for the maximum possible gain in disinfection capacity by optimising the flow. The modelled plug flow approached this maximum disinfection level well and was the most efficient system. At higher decay coefficients, the reference contactor was also more efficient, because then all the ozone is consumed in the first compartment where the plug flow was better resembled. At smaller ozone decay coefficients, the reference contactor showed a less effective disinfection and a large increase in microbial inactivation could be achieved by optimising the hydraulics. As a result, the effect of the geometrical adjustments was more pronounced at these lower decay coefficients, because in that case all the compartments contained (dissolved) ozone, so the improved hydraulics in all the compartments contributed to the gain in disinfection. An increase in log inactivation up to 50% was obtained for the smaller decay coefficients.

The inactivation rate differs per type of micro-organism; a sensitivity analysis with respect to the factor $k_{\mu} C_0$ was therefore conducted. For a range of $k_{\mu} C_0$, the log inactivation levels are plotted in Figure 4.12(b). The range of inactivation coefficients used here corresponds to micro-organisms such as *Cryptosporidium* and *Giardia* (USEPA, 1991; Rakness et al., 2005). Because the inactivation is an exponential function of the inactivation rate constant and CT value, differences in CT value become more pronounced when the inactivation rate constant increases. At higher inactivation rate constants, therefore, disinfection levels are more sensitive to hydraulic changes, i.e., improving the hydraulics by changing the geometry becomes more effective.

To judge whether flow optimisations have a significant effect on the disinfection capacity, the maximum log removal was plotted against the actual log removal in Figure 4.12(c). The maximum log removal contains all the initial information about the ozone disinfection (decay coefficient, inactivation rate constant, initial ozone concentration and hydraulic residence time, eq. 2.19). The figure was constructed by calculating the log removal from the particle tracks for a range of decay coefficients, inactivation rates and initial ozone concentrations. The small noise was caused by the variation in decay coefficient, because the CT value distribution differed for each decay coefficient due to the differences in ozone concentrations in the contactor. Therefore, different combinations of inactivation coefficients and decay coefficients could lead to similar maximum log removal values, whereas the actual log removal as calculated by the CFD model might differ, since the CT value distributions were different. Despite these variations, the effect of hydraulics and kinetics are summarised in this figure, which can be used in the following manner: given the conditions of an ozone installation, a maximum log removal for a certain micro-organism can be calculated, from which the actual log removal can be read from the figure. This determines whether a hydraulic optimisation leads to a significant increase in microbial safety. For a small maximum log removal (smaller than 1), this will not be the case. But there is a distinct trend that the effect of optimising the hydraulics is larger when higher log inactivation levels are desired. In practice, the required log inactivation usually lies between 1 and 4. The figure shows that optimising the hydraulics for this range of inactivation levels has a significant effect.

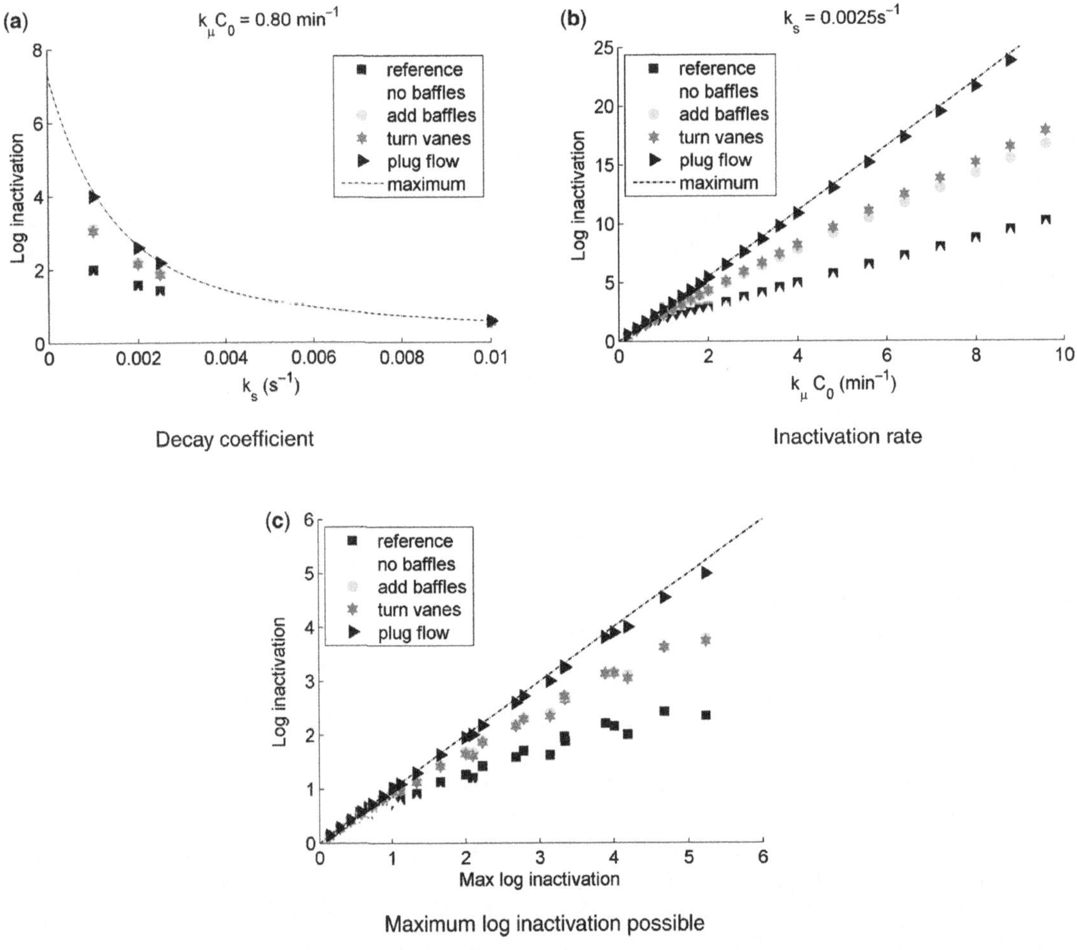

Figure 4.12 Sensitivity of the log inactivation with respect to different kinetic parameters.

4.7 SHORT-CIRCUITING

The suboptimal performance is explained by the development of distinct flow patterns of large recirculation zones and short-circuits inside a compartment. A good contactor design therefore requires a minimisation of the short-circuits. The geometric modifications to the ozone contactor are again shown in Figure 4.13. The trajectories of the particle with the lowest CT value and the particle with a mean CT value are shown. In contrast with the lowest CT value, the mean CT value did not show much differences between the contactors (Table 4.3). The particle with the lowest CT value followed the shortest pathway through the system. By investigating these trajectories, the weak points in the contactor as well as the effect of geometric modifications become apparent. The paths of short-circuiting in the reference contactor followed the outer walls and were forced to move upward or downward in the neighbourhood of a baffle. Due to the narrow openings under/above the baffles, the short-circuiting region remained small. Velocities were therefore high in the short-circuit region, so the CT value remained small here. The paths of short-circuiting in the contactor without baffles was very short, almost a straight line to the outlet, which explains the low efficacy of this contactor. Additional baffles obstructed the shortest paths from the reference contactor, so that the particles with the lowest CT values were forced to take a longer path and received a higher CT value. The turning vanes above and below the baffles prevented the development of the distinct pattern of a short-circuiting region and a large recirculation zone in each compartment. As a result, the short-circuit flow region became wider and velocities were lower, so that the minimal CT values were increased.

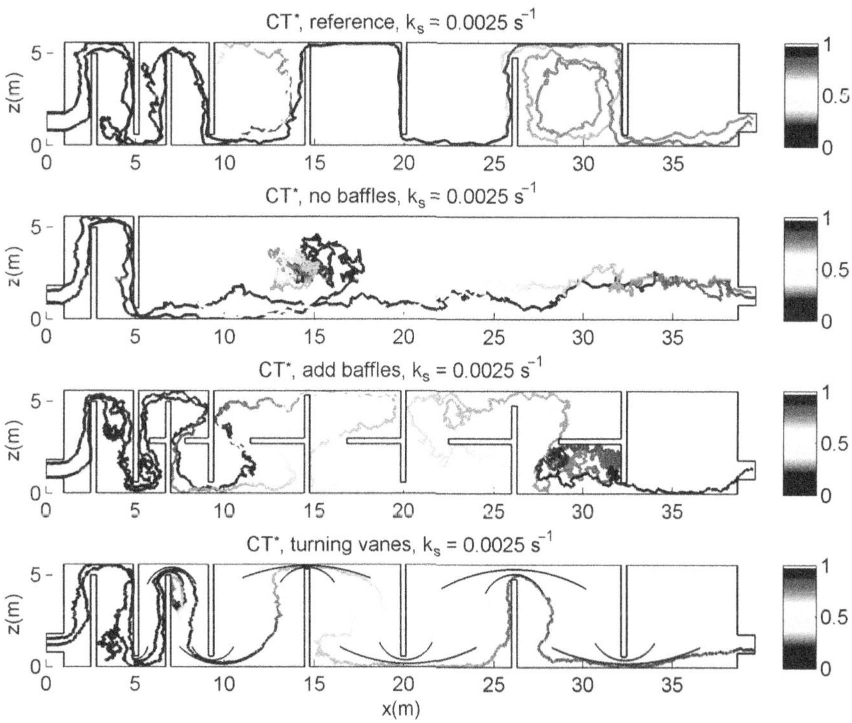

Figure 4.13 Particle tracks that represent the movements of micro-organisms inside the different contactors. The CT* is indicated by the colour. For each contactor variant, the two tracks with the lowest and mean CT value are provided.

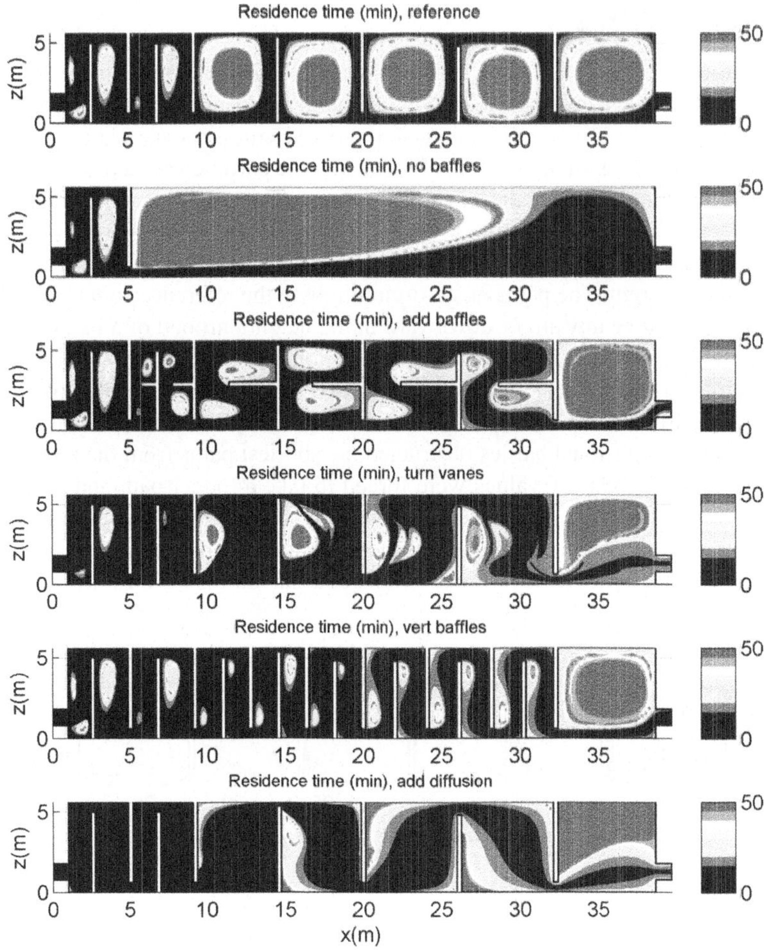

Figure 4.14 Quick scan advection residence times for the velocity field in ozone contactors. The regions with high residence times represents the dead zones, whereas the regions with low residence times represent the short-circuits.

4.7.1 Quick scan

Inspired by the importance of the short-circuiting, a computational efficient way to scan the hydraulics was investigated. By this method, the short-circuits are easily calculated, which provides a quick scan for the contactor performance. Therefore, a concentration is solved for an advection-decay equation (without diffusion), given by:

$$\frac{\partial C}{\partial t} + u\frac{\partial C}{\partial x} + k_s C = 0, \tag{4.1}$$

where the flow field is given by u. An analytical solution can be obtained for this equation, which defines the residence time (for advection only) by considering the material derivative form of eq. 4.1, and integrating over time:

$$\frac{DC}{Dt} = -k_s C, \tag{4.2}$$

$$C = C_0 \exp\left(-k_s t\right), \tag{4.3}$$

$$t = -\frac{1}{k_s} \ln\left(\frac{C}{C_0}\right). \tag{4.4}$$

The decay coefficient k_s can be chosen as an arbitrary small number. The advantage of this method is that the residence times (for only advection) are obtained quickly over the complete installation. This results in a quick overview of the weak points of the velocity field: the locations of stagnant zones and short-circuits. This method provides a worst case scenario, diffusion will improve the hydraulics, because it mixes the water in the stagnant zones (or short-circuits) with the main flow. In Figure 4.14, the results are shown for different contactor geometries. Some other variants were considered as well: additional vertical baffles and a contactor with additional viscosity (equal to $5 \cdot 10^{-2}$ m^2/s). The additional viscosity can be established by placing small elements in the contactor (for example wire-framed balls) that increase the mixing. The geometric modifications improved the velocity field clearly, the large residence times caused by the stagnant zones in the compartments were reduced.

4.8 CONCLUSIONS

Problems with disinfection by-products and the demand for microbial safety require an accurate disinfection estimation. Using a CFD model that is validated by experimental residence time distributions, the required variables for different disinfection methods were determined. The particle tracking method, which accounts for the complex flow and its turbulent fluctuations, approaches the movements of individual micro-organisms in the system most closely. Therefore, other disinfection calculation methods were compared to the particle tracking method. Small residence times and corresponding low ozone exposures (CT value), caused by short-circuit flows, were mainly responsible for a poor disinfection performance. Methods that only incorporate a residence time distribution may give inaccurate results, because the actual mixing state of micro pollutants with ozone is not incorporated in these methods. The full distribution of CT values needs to be considered in the disinfection calculation. Also, methods that do not incorporate the short-circuit flows in a correct manner, like the segregated flow analysis or Eulerian approach that calculates the mean CT value, will overestimate the disinfection capacity. Moreover, using these methods for optimisation will predict the effect of hydraulic improvements inaccurately. A particle tracking method and direct Eulerian inactivation (that solves the scalar transport equation for the number of micro-organisms directly) incorporates these details and will therefore predict disinfection more accurately.

Modifications proposed for an existing ozone contactor, such as additional horizontal baffles or turning vanes, increased the disinfection performance significantly, whereas the mean CT remained the same. As a result, the bromate production, which is proportional to the mean CT value (van der Helm et al., 2007), will not be increased by the hydraulic improvements. Alternatively, the ozone injection can be lowered to obtain similar disinfection results, whereas bromate production is reduced. The gain in disinfection performance depends on the kinetics, represented by the slow ozone decay coefficient and microbial inactivation rate. Optimising the hydraulics has less effect at small decay coefficients or for micro-organisms that are very resistant to ozone (small inactivation rate constants). But in the range of required log removal for micro-organisms encountered in practice, improving the hydraulics has a significant effect on the disinfection.

Chapter 5

Single cross-flow UV lamp systems*

The hydraulics around a single UV lamp are investigated in this chapter. Experiments were conducted to investigate measures that reduce the wake behind a lamp. Using the modelling tools from Chapter 2 and 3, a comparison is made between a LES model and a k-ε model for the velocities, UV doses, disinfection and oxidation performance.

5.1 INTRODUCTION

Large UV systems used in water treatment facilities usually contain a number of cross-flow UV lamps. These lamps lead to a disturbance in the flow, and cause a complex turbulent flow pattern - a major challenge for present day CFD models (Breuer, 2000). Flow separation occurs at the downstream side of the lamp, where a recirculation zone or wake region develops. The UV radiation emitted at the downstream part of the UV lamp is less effective, because it irradiates the water in the recirculation cell that has already received a high dose. Therefore, the flow around a single UV lamp was optimised by experimental and numerical work. Small changes to the turbulence around the lamp and changes to the flow in the boundary layer of the lamp, which could not be captured by the CFD models used here, were investigated experimentally. The effect of baffles in the neighbourhood of the lamps, which are a well-known measure to improve UV systems, were investigated numerically.

The size of the recirculation cell downstream of the lamp depends on the position where the flow separates from the cylinder surface. The point of separation at the cylinder surface occurs at the point where the fluid in the boundary layer cannot overcome the adverse pressure gradient, generated at the rear of the cylinder. So, the boundary layer at the cylinder surface largely determines the separation, which is influenced by the turbulent properties of the flow. At high Reynolds numbers (of the order of $2 \cdot 10^5$, scaled on the cylinder diameter), a transition from a laminar to a turbulent boundary layer occurs. As a consequence, additional

* Parts of the research described in this chapter are published in: Wols, B.A., Uijttewaal, W.S.J., Rietveld, L.C., Hofman, J.A.M.H and van Dijk, J.C. (2009). Hydraulic optimization of a single UV lamp placed perpendicular to the flow direction by experimental and numerical techniques. In *IO3A 19th World Congress & Exhibition*, Tokyo, Japan.
Wols, B.A., Uijttewaal, W.S.J., Hofman, J.A.M.H., Rietveld, L.C. and van Dijk, J.C., The weaknesses of a k-ε model compared to a LES model for the prediction of UV dose distributions and disinfection. Chemical Engineering Journal, 162(2): 528–536, http://dx.doi.org/10.1016/j.cej.2010.05.055. The data can be accessed through http://dx.doi.org/10.4121/uuid:c1ac7344-1419-4398-ba13-c757551c303f

energy is fed to the boundary layer by turbulent momentum exchange, so that the capability to overcome the adverse pressure gradient is increased. At this transition, the boundary layer is still laminar, but immediately behind separation there is a transition to turbulent conditions followed by a turbulent reattachment point on the cylinder. As a result, the point of separation is shifted downstream (Niemann and Hölscher, 1990). This effect, which occurs at a narrow range of Reynolds numbers, is called the drag crisis. Consequently, the recirculation cell is contracted and the drag coefficient is reduced. At higher Reynolds numbers, the point of separation shifts again a little upstream. By increasing the roughness of the lamp surface the drag crisis is advanced to a lower Reynolds number (Achenbach, 1971). Another measure that may reduce the wake size is placing a grid upstream of the cylinder to increase the turbulent intensity upstream of the lamp, which affects the boundary layer at the lamp surface. This may also result in a reduction of the recirculation zone (Niemann and Hölscher, 1990). The above mentioned hydraulic measures were evaluated by measuring the velocity profiles and comparing them to the measured velocity profiles of a reference cylinder.

The non-trivial turbulent flow pattern in cross-flow UV lamp system is a challenge for CFD models. The flow fields predicted by k-ε models are often inaccurate, especially around complex geometries (Rodi, 1997; Lübcke et al., 2001). The measured velocity fields around the reference cylinder are therefore used for the validation of the CFD models. By using a large-eddy simulation (LES) that resolves most of the turbulent eddies, the flow fields around cylinders are predicted more precisely (Breuer, 2000). Consequently, predictions for the disinfection will be different. Liu et al. (2007) investigated several turbulence models in combination with the RANS equations, and compared the computed velocity fields to experimental data in a UV reactor. They found an under prediction of the wake size by the standard k-ε model. Other turbulence models also deviated from measured velocity fields; the k-ω model model showed a better prediction of the wake size, but the far wake region was predicted wrongly. Liu et al. (2007) showed that the Reynolds stress model showed the best results, probably because it accounts for the anisotropic diffusion. Despite the differences in averaged flow fields, the differences in disinfection levels were small (within 5%). Some authors included the dynamic behaviour of the vortex shedding by means of an unsteady k-ε turbulence model and found only small differences with the stationary approach (Lyn and Blatchey III, 2005; Munoz et al., 2007). However, they still used the k-ε model, so not all the relevant turbulent length scales were resolved.

Therefore, we used the LES model which represent the physical processes more precisley and compared it to the RANS approach with a standard k-ε turbulence model. An assessment was made of the effect of both modelling approaches (LES model and k-ε model) on the residence time distributions, dose distributions and disinfection efficacy for some simplified UV systems. A cross-flow system with a single lamp is considered, focusing on the detailed flow field around a single UV lamp. It was expected that the differences between the modelling approaches were larger for a cross-flow system than for an annular systems, because the cross-flow systems exhibit a more complex turbulent flow. For example, Wright and Hargreaves (2001) investigated several turbulence models (k-ε model and Reynolds stress model) for an annular UV system, but found no differences for the predicted velocity fields and UV doses. In addition, configurations with baffles in the neighbourhood of a single lamp were considered. Baffles are a commonly used construction to optimise the design of UV reactors. By considering a small part of a full-scale UV reactor in detail, such as the flow field around a single UV lamp, the reliability of different CFD models were clarified further. The aim was to assess the accuracy of the available modelling techniques with respect to the flow fields, and how the inaccuracies in the calculated flow fields influenced the calculated disinfection.

The hydraulic measurements for the various cross-flow UV lamps are discussed in §5.2. The validation of the flow fields for the cylindrical lamp by the different modelling approaches is described in §5.3. These models were assessed for various baffle configurations in the neighbourhood of the cross-flow UV lamp with respect to the dose distribution, disinfection and oxidation efficacy in §5.4.1.

5.2 EXPERIMENTS OF VARIOUS UV LAMP SHAPES

5.2.1 Setup

The measurements were conducted in a long flume (length of 14 m, width of 0.4 m, height of 0.4 m), where flow rates could be applied up to 80 l/s. The range of Reynolds numbers (based upon the cylinder diameter) that could be covered was between 6000 and 30 000. The sides of the flume consisted of glass walls which provide optical access for flow visualisation techniques as well as laser Doppler anemometry (LDA). A number of changes with respect to the design of a single cross-flow UV lamp were considered:

- a normal cylindrical lamp was chosen as a reference (case I),
- the roughness of the lamp surface was increased (case II),
- the cylindrical lamp was replaced by a wing-shaped lamp (case III),
- a grid was placed upstream of the cylindrical lamp (case IV).

A PVC tube with a diameter of 0.05 m was used to represent the UV lamp (Figures 5.1 and 5.2). For the rough lamp (case II), sand with a diameter ranging from 2.5 mm to 4 mm was glued onto the PVC tube. These roughness elements were chosen such that the critical Reynolds number occurred at the highest flow rate (Achenbach and Heinecke, 1981). Half of a PVC tube was used for the upstream part of the wing-shaped lamp (case III). The downstream part consisted of two cylindrical parts with a radius of 0.1405 m, so that the length of the downstream part was 0.08 m. For case IV, a grid was placed upstream at a distance of 0.05 m (one diameter of the cylinder). The grid spacing was 5 mm, ten times smaller than the diameter of the cylinder, so that the turbulent length scales created at the grid were one order of magnitude smaller than the cylinder diameter.

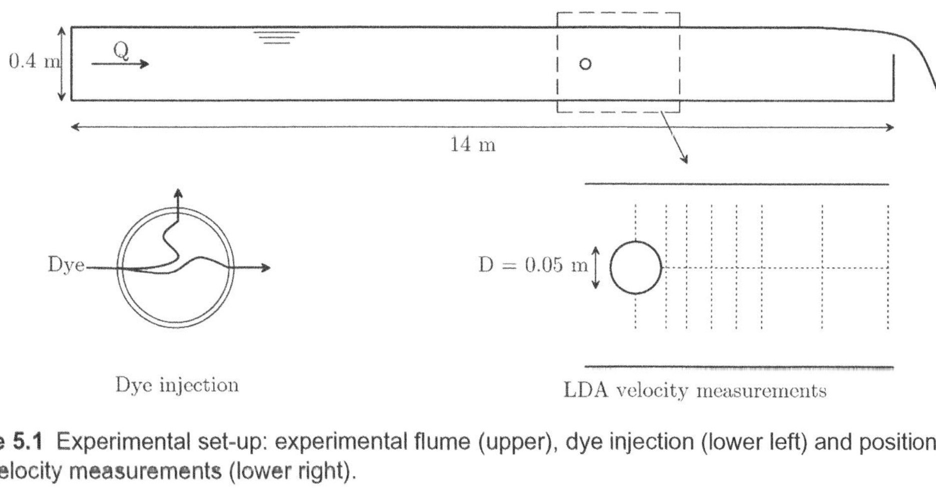

Figure 5.1 Experimental set-up: experimental flume (upper), dye injection (lower left) and positions of LDA velocity measurements (lower right).

Dye (potassium permanganate, $KMnO_4$) was injected into the flow to visualise the transport and mixing characteristics. The dye images provided a qualitative overview of flow properties. The vortices originating from the lamp as well as the wake region of the lamp were observed. The injection occurred at the top of the lamp or at the rear of the lamp (Figure 5.1). A standard video camera (3 MP, 25 fps) recorded the flow patterns.

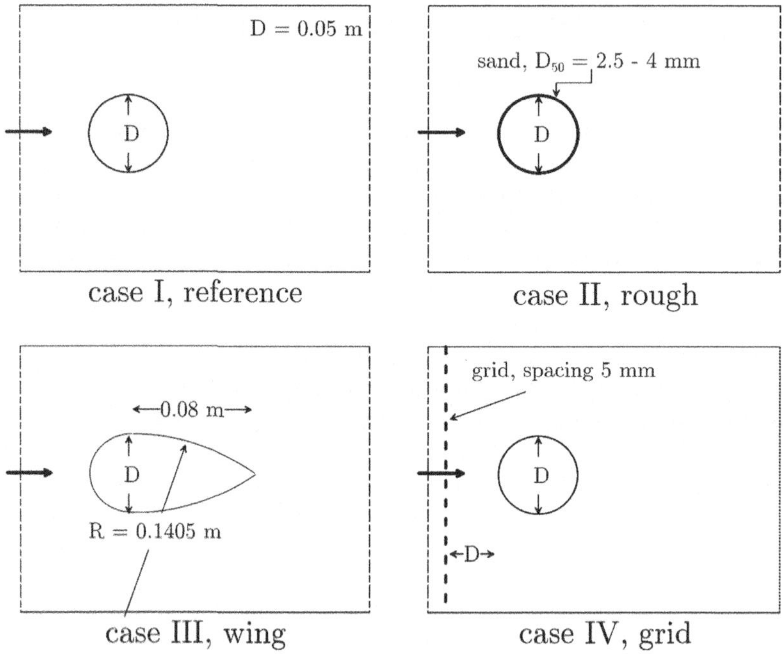

Figure 5.2 Overview of the hydraulic changes applied to the UV lamp.

The velocities and turbulent intensities were measured by means of a LDA (Ruck, 1987), which measures the instantaneous velocity at a single point in the water. Two laser beams (for two velocity directions) and a reference beam entered the water mass and were scattered by small particles in the water. Seeding with particles was not necessary, because the water contained sufficient micro particles (small enough to move entirely with the flow) to receive a steady signal. Due to the Doppler shift, the frequency of the scattered laser light changed, defining the velocity of the water in each direction. Instantaneous velocity data was captured at a sampling rate of 100 Hz for a duration of 3 minutes from which the mean velocities, Reynolds stresses (normal stresses $\overline{u'u'}$ and $\overline{w'w'}$, and shear stresses $\overline{u'w'}$), turbulent kinetic energies and power spectra were calculated. The turbulent kinetic energies (*TKE*) were calculated from the variances of the velocities:

$$TKE = \frac{3}{4}\left(\overline{(u')^2} + \overline{(w')^2}\right), \tag{5.1}$$

where u' and w' are the instantaneous velocity fluctuations (m/s). It was assumed that the third velocity direction, which was not measured, had an equal contribution to the turbulent kinetic energy as the other velocities (assuming isotropy). Adams and Rodi (1990) used a similar equation to calculate the *TKE*. The instantaneous velocities were measured in the middle of the flume at several vertical profiles downstream of the lamp (Figure 5.1).

The dynamic behaviour of the flow around the cylinders was observed by the vortex shedding frequency, which is the frequency at which the vortices are created at the rear of the cylinder. The frequency f (Hz) of the vortex shedding is characterised by the Strouhal number, defined as the dimensionless vortex shedding frequency:

$$St = \frac{fD}{u_m}, \tag{5.2}$$

where D is the diameter of the cylinder (m) and u_m the mean velocity upstream of the cylinder (m/s). The Strouhal numbers were determined from the energy spectrum of the velocity signal at each measuring point. Due to vortex shedding, there was a peak in the energy spectrum at the vortex shedding frequency. The frequency where the peak occurred was used in eq. 5.2 to calculate the Strouhal number. The vortex shedding frequency is directly related to the length of consecutive vortices, which can be related to the wake size. Since a higher vortex shedding frequency means shorter vortices and therefore a shorter wake size, the Strouhal number is in some way inversely proportional to the wake size.

5.2.2 Results of flow measurements

Dye visualisation

The dye movies gave a qualitative insight into the flow and mixing properties around the lamp. In Figure 5.3 pictures for the different cases are shown where the dye was injected at the downstream part of the cylinder ($Re = 6000$). The dye trapped inside the recirculation zone at the downstream part of the lamp is clearly visible. Also, the formation of a von Karman vortex street is visible: large vortices were shed from the top and bottom of the lamp. These large vortices gradually spread out over the complete depth of the channel and stretched out over a length of several times the channel depth. As a result, the mixing downstream of the lamp was largely increased. While the vortices moved further downstream, the energy of the large vortices was transferred to the smaller vortices by the energy cascade and eventually dissipated by viscosity.

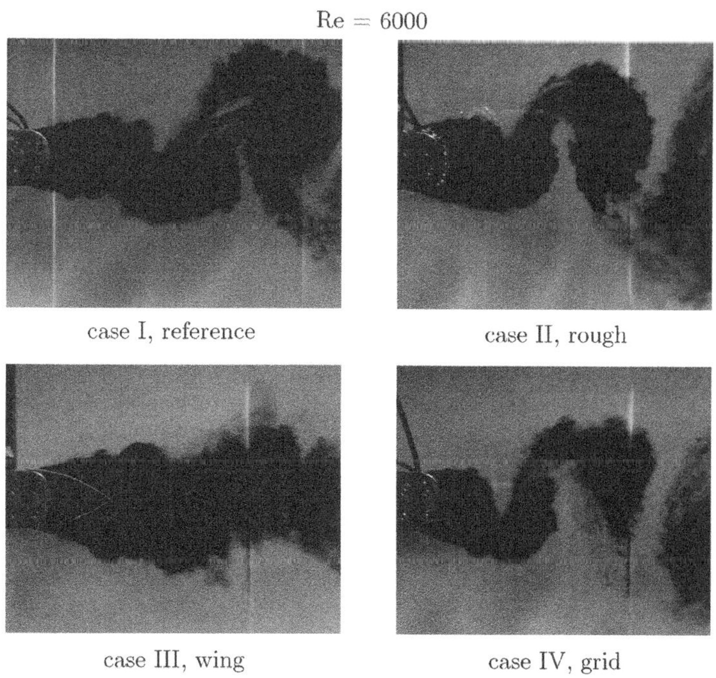

Figure 5.3 Pictures after dye injection at the downstream part of the UV lamp, showing the recirculation zone downstream of the lamp and the large-scale vortices. A video can be found at: http://dx.doi.org/10.4121/uuid:954f0e25-aa39-454d-a0fa-656f36bf72d1

The lamp with increased roughness and the lamp with an upstream grid showed a similar flow pattern as the reference cylinder. The wing-shaped lamp did not produce such distinct vortices, but still a large wake region filled with dye is observed. This indicates that a large recirculation zone as well as a large mixing layer had developed, despite the streamlined shape of the wing. In Figure 5.4, the flow around the wing shape is shown as a function of Reynolds number. While increasing the Reynolds numbers, the flow became more streamlined and remained better attached to the lamp surface. From $Re = 24\ 000$ and higher, the recirculation zone had disappeared and the vortices remained small.

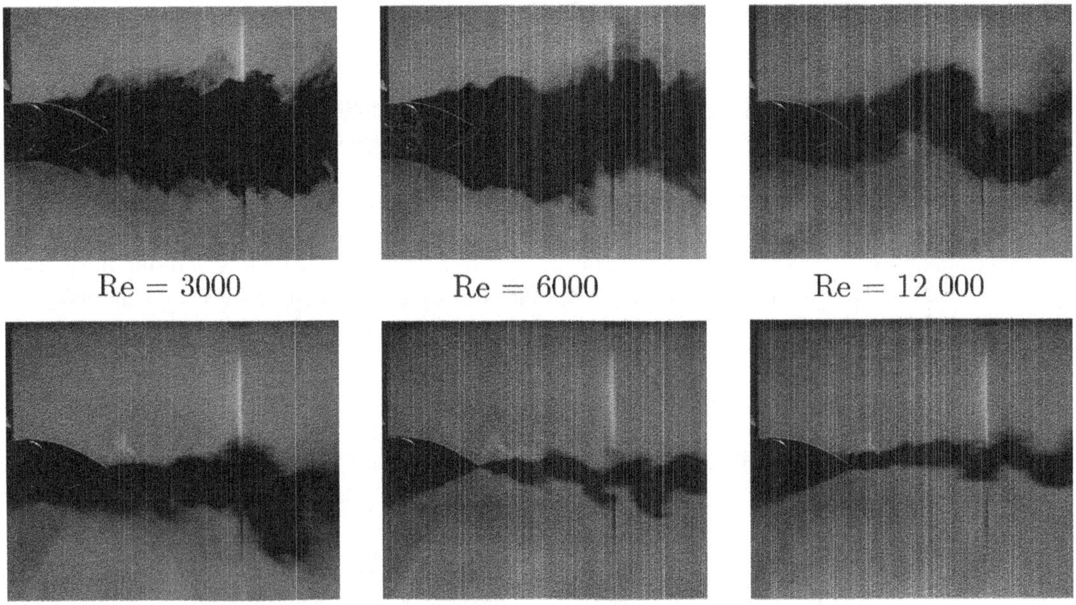

Re = 3000 Re = 6000 Re = 12 000

Re = 18 000 Re = 24 000 Re = 30 000

Figure 5.4 Dye pictures of the wing-shaped UV lamp. The flow remains attached to the cylinder surface at higher Reynolds numbers.

Velocity fields

An overview of the LDA measurements is given in Table 5.1. They were conducted for two flow rates, 16 l/s and 80 l/s, corresponding to $Re = 6000$ and $Re = 30\ 000$, respectively. The LDA measurements are summarised in Figure 5.5, where the time-averaged velocity vectors for all cases are presented. The flow was forced to move around the lamp, resulting in an upward flow above the lamp and a downward flow below the lamp. A main flow region was distinguished, which was hardly influenced by the blocking effect of the lamp in the flow. A shear layer existed between the main flow and the wake, which connected the high velocity region with the wake region. The wake region is clearly visible downstream of the lamp. Two large recirculation cells (above and below the centreline) were formed that rotate in opposite direction.

The red dots indicate the boundaries of the wake region, which were defined as the positions where the volume of water flowing in forward (streamwise) direction equalled the volume of water flowing in backward direction. They were calculated from the velocity profile (in streamwise direction) at each measured line. In Table 5.1, the normalised length (L/D) of the recirculation cell is presented, measured as the distance between the rear of the cylinder and the point of zero streamwise velocity (at the middle axis of the cylinder, $z = 0$ cm). The Reynolds number had a large influence on the wake size: a larger wake was formed at lower Reynolds numbers, which is in correspondence with other observations from literature (Breuer, 2000). The lamp with a grid in front clearly showed a large improvement with respect to the wake size for both Reynolds numbers. The increased turbulent intensity created by the grid resulted in a smaller wake. The sand elements glued on the cylinder did not have the desired effect to reduce the wake size. At the lowest Reynolds number the differences with respect to the reference case were small. However, at the highest Reynolds number the wake size was considerably increased (from 1.12D to 1.79D). A reduction was expected because the drag crisis would be reached at this range of Reynolds numbers for this roughness (Achenbach and Heinecke, 1981), which was apparently not the case. By increasing the roughness, however, the critical Reynolds number range is shrinking, and also the reduction in drag is decreasing (Achenbach and Heinecke, 1981). Also, the large roughness elements influenced the flow outside the boundary layer of the cylinder, which could influence the wake size. Moreover, the effective diameter is increased by the roughness elements resulting in a larger wake. For the wing shape an improvement was only observed at the highest Reynolds number, where no separation occurred. The separation point was very sensitive to the Reynolds number: by a small increase of the Reynolds number the separation point shifted further downstream, which was already observed in the dye experiment. As a consequence, a large wake occurred at the lowest Reynolds number, whereas the wake had disappeared at the highest Reynolds number.

Table 5.1 Overview of wake sizes and Strouhal numbers.

Case		Flow rate (l/s)	Re (*1e3)	L/D	Strouhal (mean)	Strouhal (std)
I	reference	16.5	6.2	1.62	0.2153	0.0034
I	reference	79.5	30	1.12	0.2085	0.0023
II	rough	16.4	6.2	1.63	0.1883	0.0037
II	rough	80.1	30	1.79	0.1877	0.0021
III	wing	16.4	6.1	2.72	0.1838	0.0251
III	wing	79.8	30	–	–	–
IV	grid	16.5	6.2	1.07	0.2361	0.0034
IV	grid	79.2	30	0.857	0.2410	0.0015

For a more detailed comparison of the flow fields the velocity profiles (mean velocity in streamwise direction, u, and vertical direction, w) for all the lamp shapes are plotted in Figure 5.6. The emphasis is put on the largest Reynolds number, which is closer to the range of flow conditions used at UV reactors in practice. At the centre line of the lamp ($x = 0$ cm) the velocity profiles were almost the same. Only small differences close to the lamp surface were observed. In the near wake region ($x = 5$ cm to $x = 10$ cm), the differences between the lamp shapes were distinct. At $x = 7.5$ cm, the average streamwise velocity was still negative for the normal and rough cylinder, whereas the recirculation had disappeared for the cylinder with a grid in front. For the wing shape, no recirculation was present, only a reduction in velocity to half of the inflow velocity was observed directly downstream of the wing (at $x = 10$ cm, $z = 0$ cm). In the far field ($x = 15$ cm), the lamp shapes with the shortest wakes recovered quickest from the disturbance of the velocity profile.

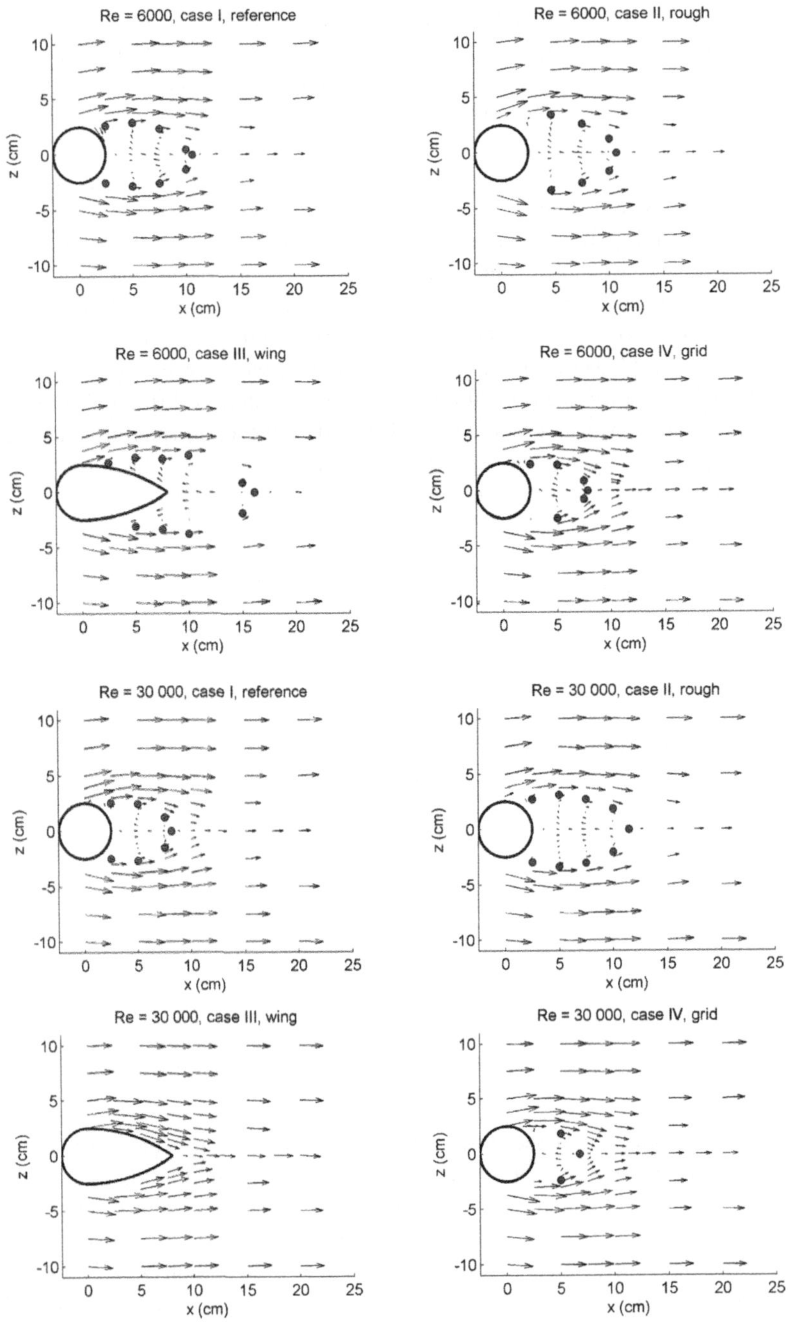

Figure 5.5 Vector plot of the measured velocities for the various cross-flow UV lamps (upper panels: *Re* = 6000, lower panels: *Re* = 30 000). Dots indicate the boundary of zero mean velocity.

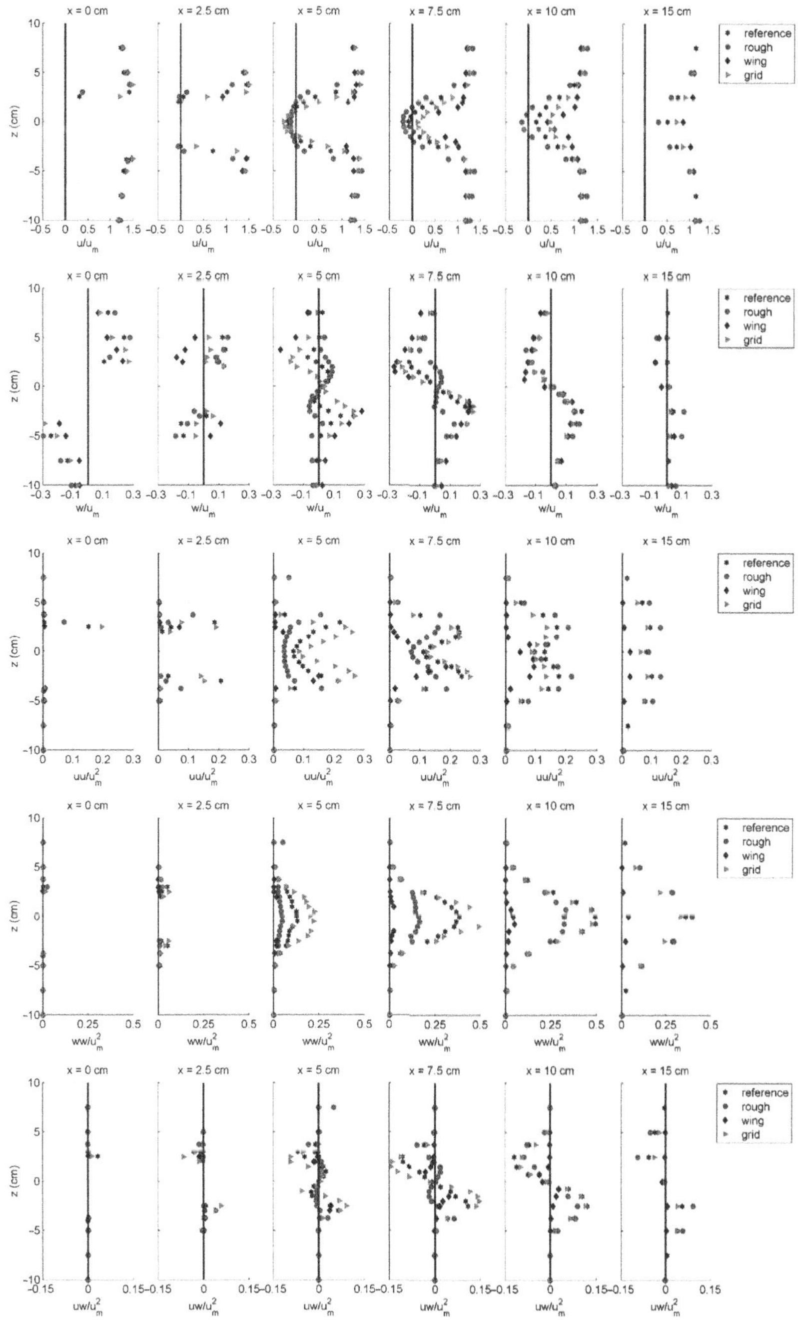

Figure 5.6 Measured velocities and Reynolds stresses at $Re = 30\,000$, mean velocity in streamwise direction (u), mean velocity in vertical direction (w), normal stress in streamwise direction ($\overline{u'u'}$), normal stress in vertical direction ($\overline{w'w'}$) and shear stress ($\overline{u'w'}$). The values are normalised with respect to the average inflow velocity (u_m).

Reynolds stresses

The Reynolds stresses show more details of the complex flow patterns as well as the mixing characteristics (Figure 5.6). The differences between the lamp configurations were related to the size of the wake region, so that the local maxima and minima of the stresses occurred at different positions, but similar patterns were observed. Observing the normal stress profile (in streamwise direction, $\overline{u'u'}$) shows that the maximum stresses occurred at the shear layer, whereas a local minimum occurred at the centre line of the cylinder ($z = 0\,\text{cm}$). For the normal stress in transverse vertical direction ($\overline{w'w'}$) the maximum value occurred at the centre line of the lamp. The vortices originating from the lamps caused an alternating upward and downward velocity, which explained the large values of these normal stresses. The shear stresses ($\overline{u'w'}$) had their maximum values outside the recirculation zone, in the middle of the shear layer between the recirculation and main flow. Further downstream, the stress profiles diffused out over the complete height, which was in accordance with the observation from the dye experiments that the vortices were spread out over the height. The turbulent energy inside these vortices was spread over the height and slowly reduced by viscosity. A general pattern was observed that the lamp configurations with shorter wakes yielded higher Reynolds stresses but over a shorter downstream distance. So, the momentum transport is higher, enabling a quicker redistribution of the velocity profile, in other words, reducing the wake length.

Strouhal numbers

The vortex shedding occurred at several positions downstream of the cylinder, where the vortex shedding frequency was determined from the measured instantaneous velocity signals. The ensemble average of these vortex shedding frequencies was taken, from which the Strouhal number was calculated (Table 5.1), based upon the mean velocity in the flume. Standard deviations of the Strouhal data points were small (of the order of 1% of the average). The Strouhal number for the reference cylinder was close to other experimentally observed Strouhal numbers, $St \approx 0.2$ (Achenbach and Heinecke, 1981). The adjustments to the cylinder had their effect on the Strouhal number: a decrease occurred for the rough cylinder, whereas an increase occurred by placing a grid upstream. The decrease in Strouhal number could be explained by the increase in effective diameter by the sand particles, which was not incorporated in the calculation for the Strouhal number (the increase of 10% of the effective diameter would give similar Strouhal numbers as for the normal cylinder). Niemann and Hölscher (1990) pointed out that changing the free stream turbulence by placing a grid upstream as well as changing the roughness may alter the Strouhal number. Achenbach and Heinecke (1981) observed an increase in Strouhal number from 0.2 to 0.25 at the drag crisis. Clearly, the sand particles glued on the cylinder did not have the desired effect: the wake size was increased and the Strouhal number was decreased, showing that the drag crisis was not reached here. For the wing shape, at the low flow rate vortices were still shed from the cylinder but with a smaller frequency. At higher Reynolds numbers, the vortex shedding disappeared, because the vortex shedding mechanism was stopped.

5.2.3 Discussion

The experiments showed the sensitivities of the wake formation behind cylinders. The interaction between the boundary layer at the surface of the cylinder and the turbulent flow is complex, and influences the separation point, and, consequently, the wake size. Changing the turbulent properties of the flow by placing an upstream grid or changing the boundary layer by adding roughness elements had changed the flow fields considerably. From the experiments, it is concluded that the most effective measures to reduce the wake size of cross-flow UV lamps are placing a grid upstream of the lamp and changing the cylindrical shape of the lamp to a wing-shaped lamp. Adding roughness elements to the cylinder surface did not have the desired effect, because the drag crisis was not reached. The effectiveness of measures that reduce the wake size depended on the Reynolds number, which was clearly observed for the wing shape. The designs may be tuned to reduce the wake size further: by adjusting the grid spacing, distance of the grid to the lamp and size of the wing.

5.3 CFD MODELLING OF THE REFERENCE CYLINDER

5.3.1 Modelling conditions

Using the measured velocity profiles of the reference cylinder, both CFD models (LES model and k-ε model) were validated. A flow domain with dimensions $0.4 \times 0.8 \times 0.4\,\text{m}^3$ (using a similar width and height as the experimental flume) was used for the CFD calculations (Figure 5.7). For a RANS model with k-ε turbulence modelling two meshes were used: a mesh with 41 228 elements and one with 132 228 elements. The following boundary conditions were imposed: a velocity at the inlet, a pressure at the outlet, velocities corresponding to a logarithmic wall function at the walls and a free slip condition at the free surface. For the LES calculations three different mesh sizes were used: 1.8 million, 4.2 million and 14.2 million elements. The velocities at the boundaries were set to zero (no slip condition), whereas a velocity corresponding to a logarithmic velocity profile was imposed at the first mesh point perpendicular to the wall. At the cylinder boundary, the velocities were set to zero. At the inlet, a logarithmic velocity profile was prescribed, whereas at the outlet a zero pressure was imposed.

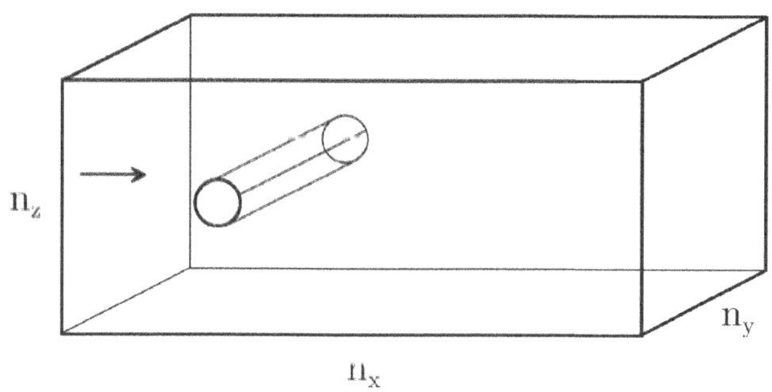

Figure 5.7 Sketch of the computational domain of the CFD models.

5.3.2 Validation of velocity profiles

The comparison between the measured velocity profiles and the CFD models is presented for $Re = 30\,000$. Similar results were obtained at the lower Reynolds number. The streamwise velocities calculated by the k-ε model are shown in Figure 5.8(a). Mesh independency was reached for 41 228 elements, since refining the mesh showed similar results. The velocity profile near the cylinder ($x = 0\,\text{cm}$) showed a good agreement with the measurements, so the upstream flow conditions are predicted correctly. However, the model predicted a very small recirculation zone (around 1 cm). As a result, the velocity profiles downstream of the cylinder were predicted wrongly by the model. Not resolving the turbulent motions and the large-scale vortex shedding resulted in a substantial under prediction of the wake. Also the increase in streamwise velocity directly above and below the wake region was not predicted by the model. Further downstream, at the far wake region ($x = 15\,\text{cm}$), the CFD model again coincided with the measurements. Also, the turbulent kinetic energies (*TKE*) were largely underestimated by the k-ε model, Figure 5.8(b). Close to the cylinder, in the near wake, the *TKE* were in the same order of magnitude as the measured kinetic energies.

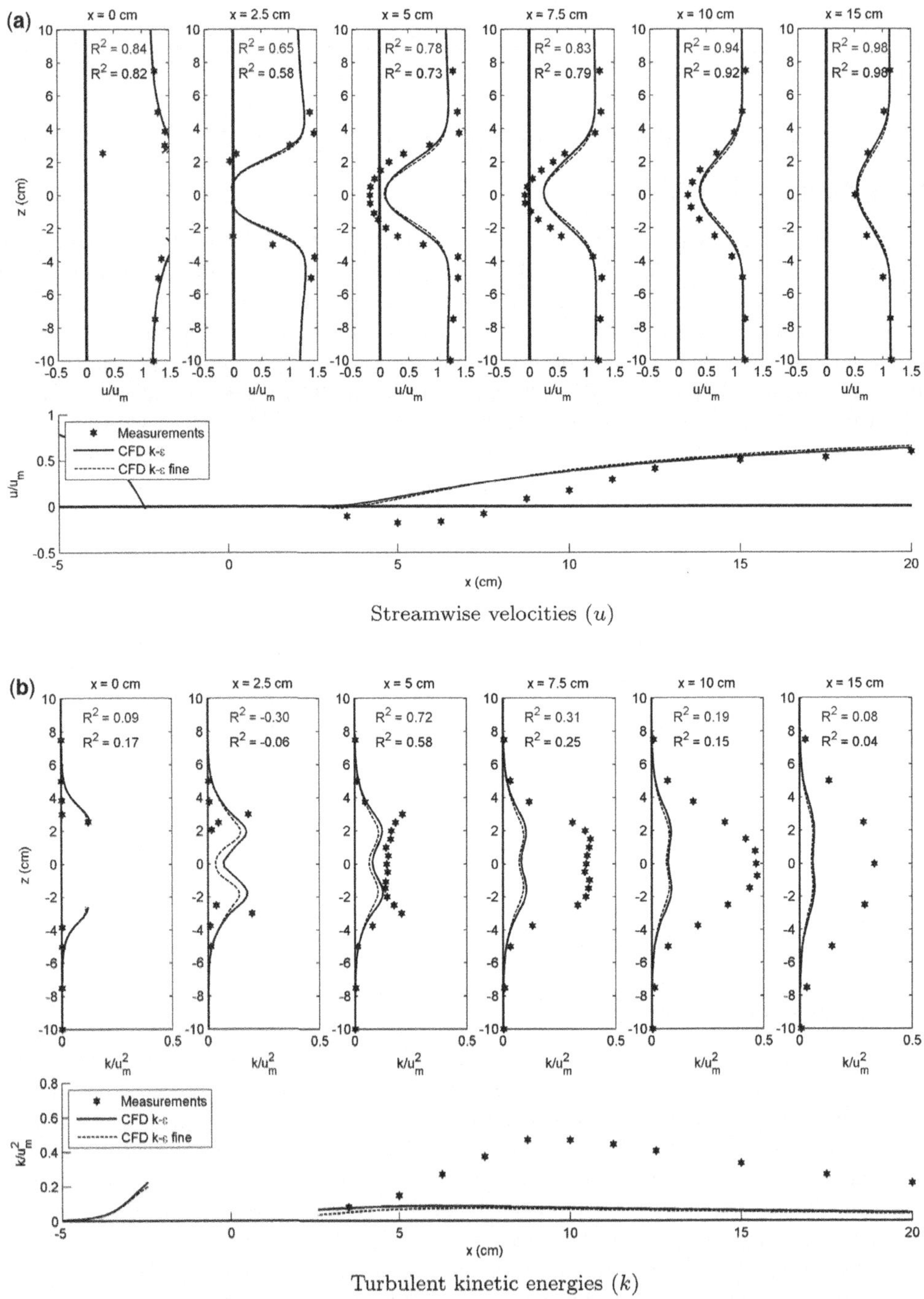

Figure 5.8 Comparison between measurements and k-ε model calculations for different mesh sizes (CFD k-ε: 41 228 elements and CFD k-ε fine: 132 228 elements). The coefficients of determination (R^2) are indicated for both mesh sizes.

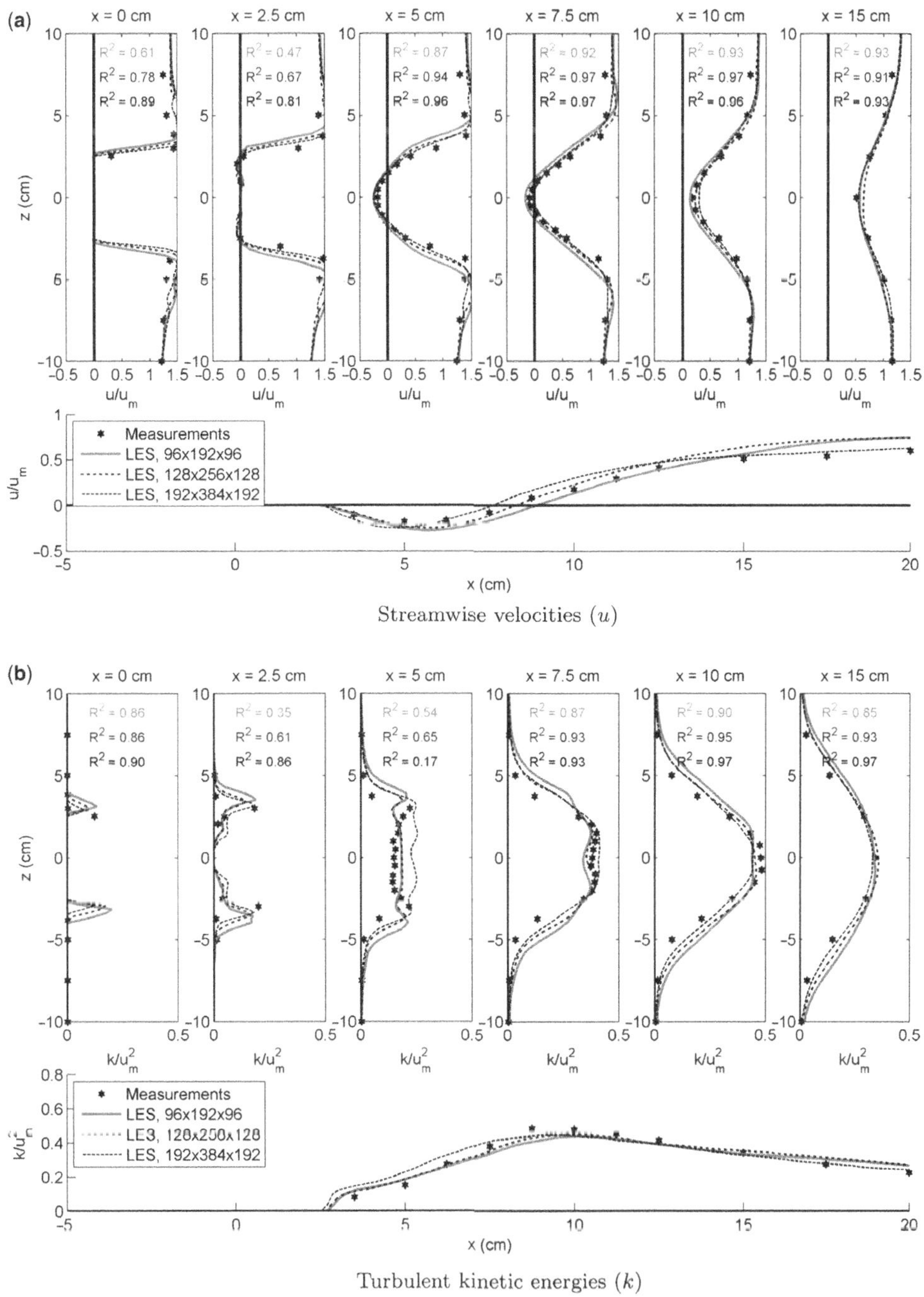

Figure 5.9 Comparison between measurements and LES model calculations for different mesh sizes. For each mesh size, the number of elements in vertical, streamwise and cross-flow direction is indicated. The coefficients of determination (R^2) are indicated for all the mesh sizes.

Further downstream, however, the CFD model did not follow the measured increase in *TKE*. Less turbulent energy was created, because the streamwise velocity differences over the height were smaller. However, the small turbulent kinetic energies also had their effect on the velocity profile: although the disturbance of the velocity profile by the cylinder was small, it stretched out over a large length. Therefore, at $x = 15$ cm, the velocity profile was again well predicted by the model.

The results from the LES model are shown for different mesh resolutions (1.8 million, 4.2 million and 14.2 million elements) in Figure 5.9(a). The solution was not completely independent of the mesh size, especially the velocity at the centre line of the cylinder showed variations with respect to the mesh size. Also, close to the cylinder, where the velocity gradients were the largest, the finer mesh matched closer with the measurements. Despite the differences, the computed results resembled the measurements well. Mesh dependency was also observed by Breuer (2000), where increasing the mesh resolution resulted in a shorter recirculation length and not necessarily a better agreement with measurements. A similar pattern is visible for the turbulent kinetic energies in Figure 5.9(b). Some small differences between the different mesh sizes were again visible; mainly the *TKE* close to the cylinder differed. But, in general, the turbulent kinetic energies predicted by the LES model agreed well with the measurements. The model predicted Strouhal numbers of $St = 0.23$ and $St = 0.21$ for $Re = 5000$ and $Re = 30\,000$, respectively. These values were close to the measured values, indicating that the dynamical vortex motions were resembled well by the LES model.

5.4 COMPARISON BETWEEN LES MODEL AND K-ε MODEL

Inspired by the differences in velocity field between the LES model and k-ε model for the reference cylinder, both models were evaluated with respect to residence time distributions, dose distributions and the effects on water quality. Since the CFD models can not resolve all the flow details of the boundary layer of the cylinder, other measures than discussed in §5.2 were considered: baffle configurations. Baffles are a well-known measure to improve UV systems, and largely influences the flow fields: recirculation zones are formed and the flow around the UV lamp is changed. So, configurations with baffles are a challenging test case for CFD models, and are relevant for the design of UV systems.

5.4.1 Configurations

A single cross-flow UV lamp in a rectangular box was chosen, where the lamp was placed at half of the height. The height and width of the box were both equal to a length of 3 times the cylinder diameter, so that the distance from the lamp surface to the outer wall was optimal for a transmittance of 80% (§2.7.2). Four different flow situations were considered, a reference situation and three baffle configurations to force the water elements to flow closer to the lamp (Figure 5.10):

- Reference, the reference situation with a single UV lamp in a box.
- Baffle upstream, a baffle was placed upstream of the UV lamp.
- Baffle middle, a baffle was placed at the middle of the UV lamp.
- Baffle downstream, a baffle was placed at the rear of the cylinder.

The differences between the baffle configurations were small, but they could have a large impact on the flow field downstream of the lamp. The parameters used for the modelling are shown in Table 5.2. The baffles were 0.025 m high, dividing the opening between the cylinder and bottom or top wall into half.

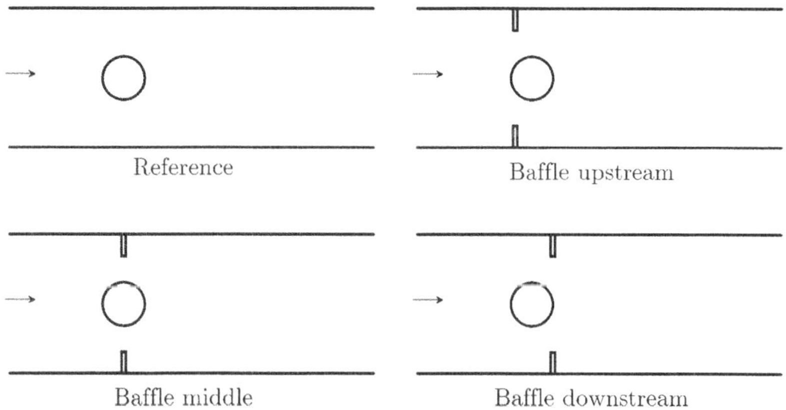

Figure 5.10 The flow around a single cylinder was investigated for different baffle configurations.

Table 5.2 Parameters used for the CFD calculations.

Flow rate (*Q*)	0.005 m^3/s
Baffle height	0.025 m
Total power UVC (*P*)	200 W
Diameter quartz sleeve	0.05 m
Thickness quartz sleeve	0.0019 m
Length of the lamp, width of domain	0.15 m
Transmittance water (T_w)	80%
Transmittance quartz (T_q)	96%
Inactivation constant *B. subtilis* (k_μ)	0.136 cm^2/mJ
Threshold inactivation *B. subtilis* (ϕ_0)	12.3 mJ/cm^2
Normalised dose (D_{scale})	132 mJ/cm^2

5.4.2 Modelling approach

For the calculated dose distribution and disinfection, the independency with respect to the mesh size, number of particles and time stepping of the particle movement was investigated. For the k-ε model, the standard conditions were defined by: a medium mesh size (41 228 elements), 4900 particles and a time step of the particle displacement of 0.025 s. These conditions were varied and the dose distribution for each variation is shown in Figure 5.11. The mean dose (D_{mean}), defined as the average of the doses of all the particles, the D_{10}, defined as the 10th percentile of the dose distribution and disinfection (*DE*) of *Bacillus subtilis* spores, calculated from the dose of each particle using the delayed Chick-Watson model (eq. 2.10, Chapter 2), are indicated as well. Considering the mesh size, a medium mesh was sufficient to obtain a converged result for the disinfection (*DE*). Regarding the number of particles, the dose distribution became smoother by increasing the number of particles. The disinfection results were converged at a number of 10 000 particles, whereas the standard condition of 4900 particles showed a deviation of 1–2% in comparison with the converged solution. To reduce computational times the standard conditions were used, allowing a small deviation. The time stepping (Δt) depended on the mesh size (Δx) and maximum velocity (u_p), presented by the CFL number (eq. 3.34). For the medium mesh size, the solution was converged at a time step of 0.025 s, which corresponded with a maximum CFL number of the particle of around 1.

The dependency with respect to the mesh size and number of particles was also considered for the LES model (Figure 5.12). Regarding the time step, the LES model used a time-varying velocity field, for which small time steps between consecutive velocity fields were needed. These time steps were also used for the particle tracking, which were small enough to obtain a accurate solution (CFL number smaller than 1). The standard conditions were: a medium mesh size ($64 \times 256 \times 64$) and 50 000 particles. Due to the fluctuations in the velocity field, instead of using a peak release of particles, the particles were injected gradually during the first 1000 time steps to obtain statistically more reliable results. By reducing the mesh size, starting from the coarse mesh ($48 \times 192 \times 48$) to the medium mesh, the predicted disinfection increased. At the medium mesh, the solution was converged; the disinfection at a medium mesh size fell within one percent accuracy with respect to the fine mesh ($96 \times 384 \times 96$). This accuracy was regarded as an acceptable choice when trading off the accuracy against computational speed. Regarding the number of particles, the solution was converged at numbers in between 10 000 and 50 000 particles. Both the LES model and the k-ε model used the standard conditions to assess the various UV systems.

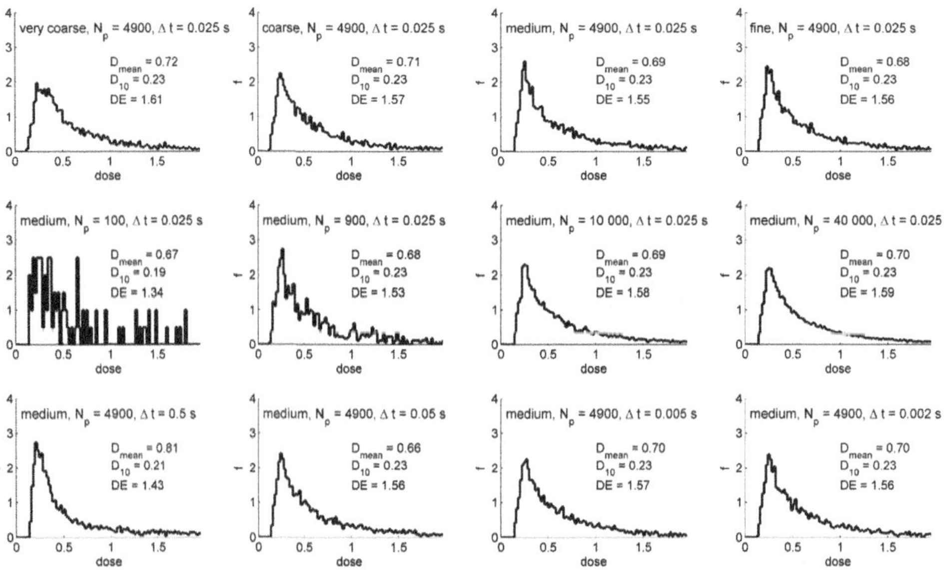

Figure 5.11 For the reference situation, normalised dose distributions and disinfection (*DE*) were calculated for different mesh sizes, numbers of particles and time steps to obtain a converged solution for the k-ε model. In the first row the variation of the mesh size is shown; in the second row variation of the number of particles; and in the third row variation of the time step.

5.4.3 Velocity fields

The velocity fields calculated by the k-ε model and LES model are depicted in Figures 5.13 and 5.14, respectively. The streamwise mean velocities (u) are shown for the different flow configurations. The velocity averaged over the inflow boundary was equal to 0.22 m/s. The water flowed around the cylinder, so that the velocity increased above and below the cylinder, whereas the velocity directly behind the cylinder decreased or even reversed in direction due to the obstacle. By placing baffles, the bulk flow was forced to move closer to the lamp. However, the bulk flow had to move through a smaller area, so that the velocity increased even further (up to 4 times the average velocity). Downstream of the baffles large recirculation zones developed. The differences between the k-ε model and LES model were large, especially for the configuration with baffles. For the reference case, differences were noticeable near the wake region. The LES model predicted a distinct region with reversed velocities behind the lamp at a length of around two times the lamp diameter. The k-ε model did not predict reversed velocities, only a reduced velocity downstream of the cylinder which stretched out over a length of a few times the lamp diameter. Also, the LES model showed that the bulk flow moved at a larger distance from the lamp (the wake region was wider).

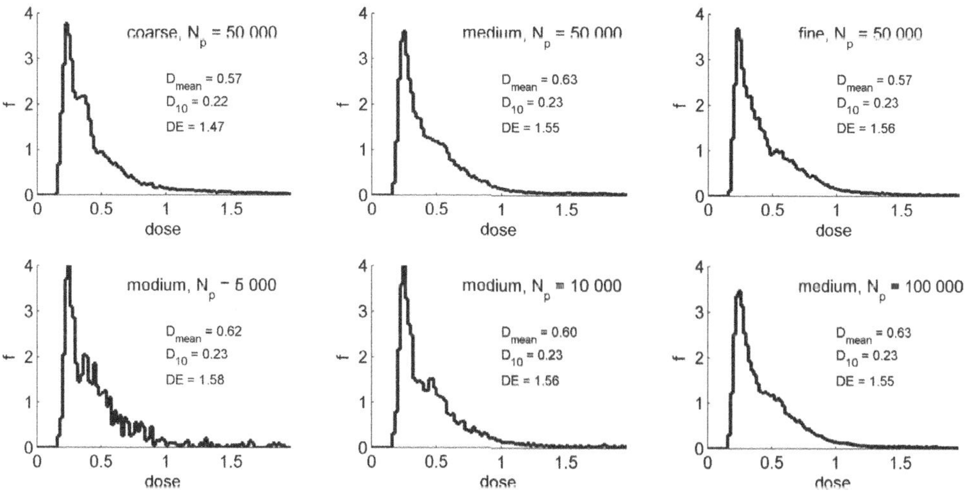

Figure 5.12 For the reference situation, normalised dose distributions and disinfection (*DE*) were calculated for different mesh sizes and numbers of particles and time steps to obtain a converged solution for the LES model. In the first row the variation of mesh size is shown; and in the second row variation of the number of particles.

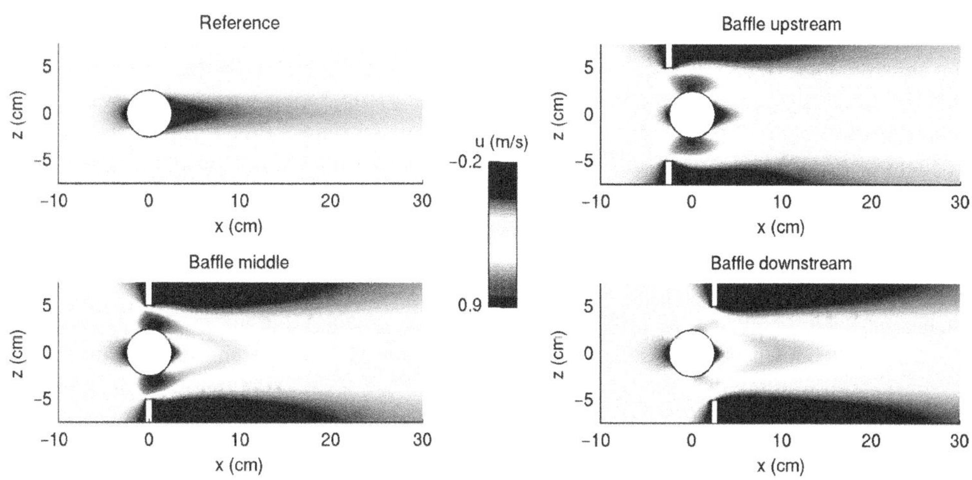

Figure 5.13 Velocity field, streamwise velocity u (m/s), calculated by the k-ε model for the different configurations.

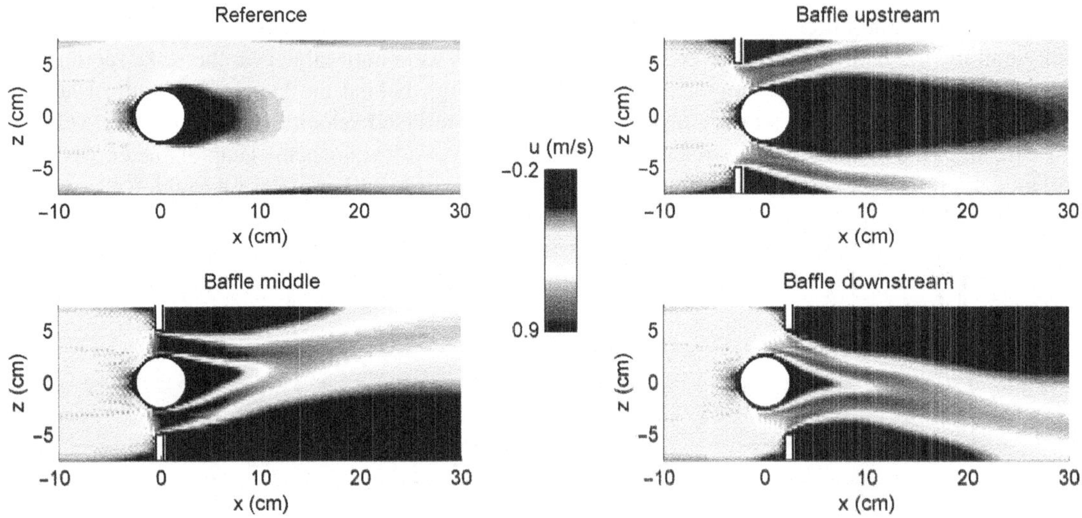

Figure 5.14 Velocity field, streamwise velocity *u* (m/s), calculated by the LES model for the different configurations.

By placing baffles above and below the cylinder, the differences between the models became more apparent. Two jets developed in between the baffles and the cylinder, one above and one below the lamp, which moved differently downstream of the cylinder in both models. The jets predicted by the LES model were more pronounced and stretched over a larger distance, whereas the jets in the k-ε model tended to quickly diffuse over the height. The differences were most pronounced for the configuration with the baffle upstream. The k-ε model predicted no reverse flows downstream of the cylinder here, and the jets had quickly merged. However, in the LES model, the jets did not merge, but moved separately towards the lower and upper wall. Also, a large wake had developed downstream of the lamp. Such a large wake may in fact be unfavourable for the disinfection performance, because the bulk flow does not enter the region with high UV intensities directly behind the lamp. For the configuration with baffles in the middle or downstream, the wakes predicted by the LES model were smaller than for the upstream baffles, because the jets had merged near the centre instead of moving to the outer walls. This showed the sensitivity of the position of the baffles to the flow field: by moving the baffle a little downstream, the direction in which the jets move had changed and the wake was considerable reduced. Small differences had a large impact on the wake formation. In contrast with the LES model, the size of the wake predicted by the k-ε model was only little influenced by the position of the baffle.

Another remarkable result of the LES model was visible for the configuration with a baffle in the middle and a baffle downstream. After the two jets had merged into a single jet, this jet had the tendency to move towards either the upper or lower wall. Due to the symmetry, there is in principal no preference for the upper or lower wall, but once the jet had moved to one side, it remained there. The so-called Coanda effect (Wille and Fernholz, 1965) explains this behaviour: due to small pressure differences the jet is moved towards one side, which increases the pressure difference, eventually resulting in the attachment of the jet to the wall. The k-ε is not capable of predicting these sensitive flow features.

5.4.4 Residence time distributions

The (cumulative) residence time distributions (RTD) are plotted in Figure 5.15. All the particles were instantaneously released in the domain at a distance of 0.15 m upstream of the cylinder. The distribution was obtained by collecting for each particle the time to cover a horizontal distance of 0.5 m. The differences between both models became also apparent in the residence time distribution. Although the differences were still small for the reference case, they clearly marked the differences in underlying velocity fields. In the LES model, the velocity gradients were larger, a higher maximum velocity occurred above and below the cylinder, which resulted in the shortest residence times (T_{10} was shorter). Also, the larger wake region downstream of the cylinder caused the longest residence times (T_{90} was longer). For the configurations with baffles, the differences became more pronounced. The LES model predicted much shorter residence times than the k-ε model. The jets predicted by the LES model were more confined and the recirculation regions were much larger than for the k-ε model. The region of the bulk flow was therefore smaller and, consequently, this region had higher velocities and corresponding lower residence times. The RTD was changed a little by the position of the baffles in the k-ε model. The LES model predicted large differences in the RTD between the baffle configurations, for example the T_{90} was considerably increased by placing the baffle further downstream. Remarkably, for all the baffle configurations the RTD became wider than for the reference configuration. If the RTD would characterise the performance of UV systems, the effect of placing baffles would be negative. In fact, the hydraulics became less optimal by placing baffles, because larger short-circuits and larger recirculation occurred. However, the spatial distribution of UV irradiance needs to be considered as well.

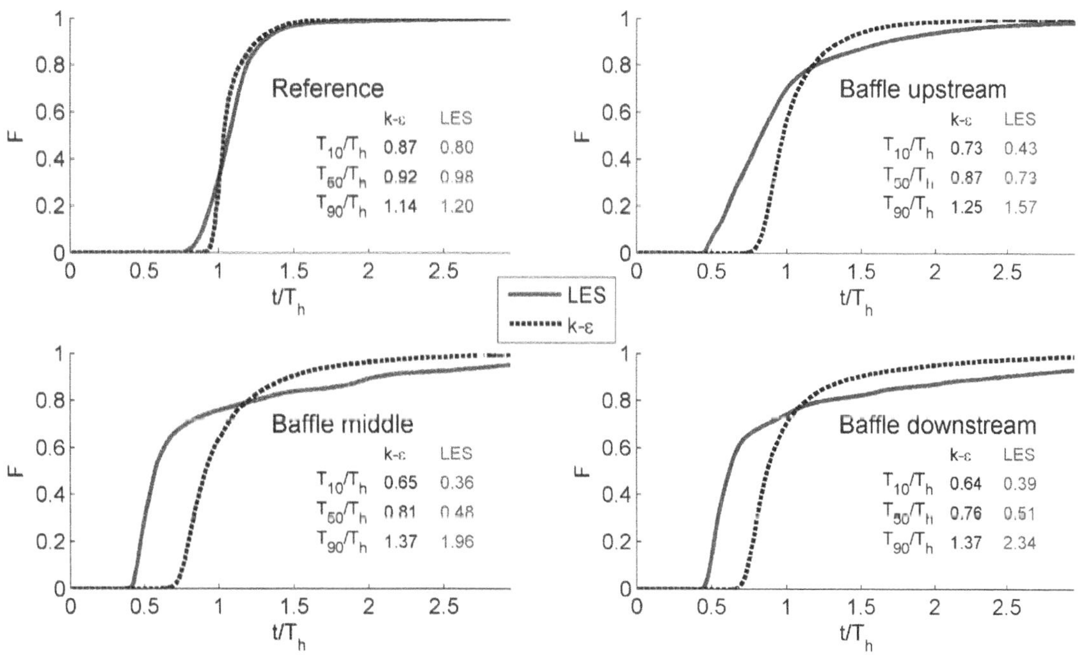

Figure 5.15 Cumulative residence time distribution calculated by the k-ε model and LES model for the various geometries. Residence times are normalised with the hydraulic residence time.

5.4.5 Dose distributions and disinfection

The UV dose distributions are shown in Figure 5.16. The mean dose, 10th percentile dose and inactivation of *Bacillus subtilis* spores are indicated as well. For the reference case, the differences between the models were small. The peak in the dose distribution was found at the same dose. Since the lower dose part in the distribution was predicted similar for both models, the disinfection predictions were also equal. The differences were large in the case of the upstream baffle. For the LES model the peak in dose distribution was found at a similar dose as for the reference situation, whereas for the k-ε model the peak shifted towards a higher dose. As a result, the k-ε model predicted a considerable increase in disinfection (from 1.55 to 2.68 log units) by placing this upstream baffle, whereas the LES model predicted a decrease in disinfection (from 1.55 to 1.45 log units). The k-ε model showed similar results for the configuration with the baffle in the middle, where the dose distribution closely matched with the upstream baffle configuration. The LES model calculated the peak in the dose distribution at a slightly higher dose than for the reference case, resulting in an increased disinfection. However, the peak was still found at a smaller dose than for the k-ε model, so that the LES model still predicted a smaller disinfection than the k-ε model. By placing the baffle further downstream, the disinfection predicted by the LES model was further increased to 2.11 log units, but not to the extent of the k-ε model prediction (2.49 log units). Here, the differences in dose distribution between the models were smaller than for the other baffle cases, but the LES model still predicted the peak at a lower dose and therefore lower disinfection levels. Where for the different baffle configurations the residence time distribution became wider than for the reference configuration, the dose distribution became narrower by placing baffles (except for the baffle upstream). The short-circuits moved close to the lamp, so that the short residence times were compensated with high irradiances. Therefore, the dose distribution better characterises the performance of a UV system than the RTD, since it incorporates the UV radiation distribution.

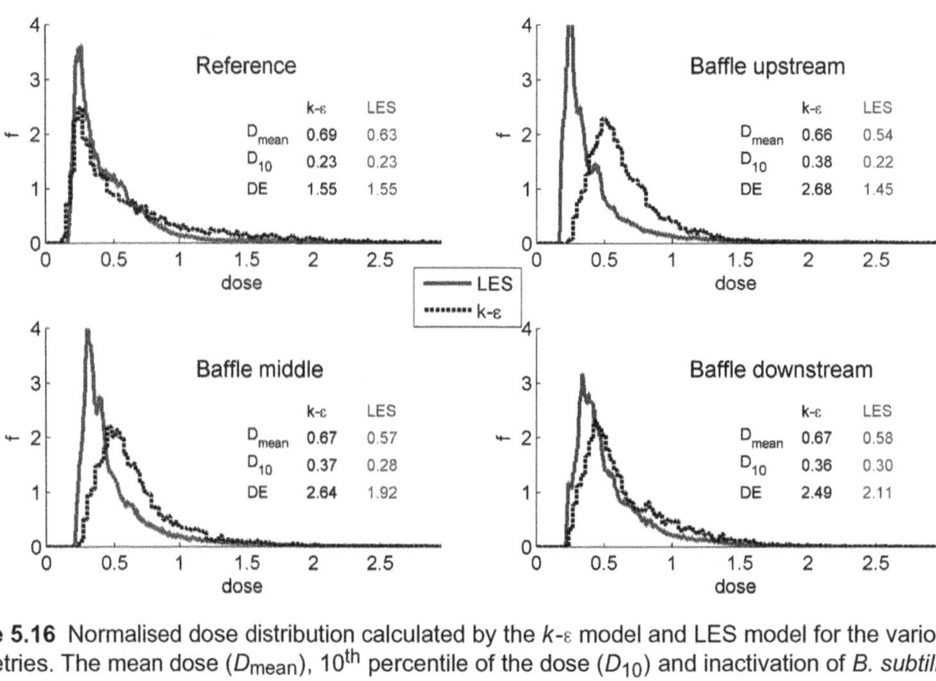

Figure 5.16 Normalised dose distribution calculated by the k-ε model and LES model for the various geometries. The mean dose (D_{mean}), 10th percentile of the dose (D_{10}) and inactivation of *B. subtilis* (DE) are indicated as well.

The disinfection predictions are summarised in Figure 5.17 and Table 5.3. The predictions were equal for both modelling approaches in the case of the reference situation. Surprisingly, the baffle configurations showed opposite patterns for both models: placing the baffle further downstream resulted in an increase in disinfection for the LES model, whereas the k-ε model predicted a decrease in disinfection. This increase in disinfection for the LES model can be explained by the size of the wake: moving the baffles downstream reduced the wake, so that the bulk flow moved through an area with higher UV intensities, resulting in higher UV doses. The best results for the k-ε model were obtained for the upstream baffle (improvement of 73%), where the LES model predicted the worst disinfection results (reduction of 6%). Placing the baffle downstream resulted in an improvement of 36% in disinfection predicted by the LES model.

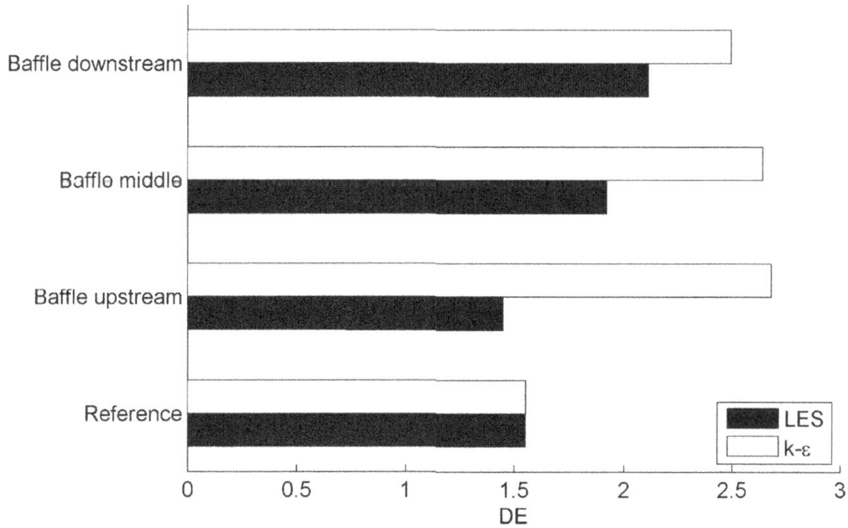

Figure 5.17 Overview of the disinfection levels, predicted by the LES model and k-ε model for the various geometries.

Table 5.3 Improvement of placing baffles with respect to the reference configuration.

Configuration	k-ε model (%)	LES model (%)
Baffle upstream	73	−6
Baffle middle	70	24
Baffle downstream	61	36

5.4.6 Oxidation

The state-of-the-art for modelling oxidation would be predicting the concentration of contaminant and peroxide simultaneously, and solving the concentration gradients induced by the differences in UV irradiances and turbulent motions. This modelling approach is followed by the LES model, where the degradation of atrazine and the reduction of peroxide is calculated simultaneously in an Eulerian framework using the equations described in Chapter 2. For the reference cylinder this approach is followed and the reduction of atrazine is computed. A monochromatic lamp with a power of 2000 W (UVC) was used to establish the high UV dose needed for oxidation, and the initial concentration of hydrogen peroxide was set to 10 mg/L. Furthermore, the parameters provided in Table 2.2 were used for the calculation of the atrazine concentration. A snapshot of the concentration of atrazine and the concentration of peroxide is shown in Figure 5.18. The largest reduction of peroxide occurred directly downstream of the lamp, where the UV irradiances were high and the residence times were long due to the recirculation cell. Nevertheless, the reduction of peroxide did not exceed 30%, and the average reduction of peroxide at the effluent was around 5%. The reduction of atrazine was also largest in the wake region, where more than 99% of the atrazine was removed. The reduction of atrazine was smallest close to the outer walls, where due to the low irradiances and high velocities only 30% of the atrazine was removed. Further downstream regions with different concentrations of atrazine were mixed, so that on average a reduction of around 65% was established.

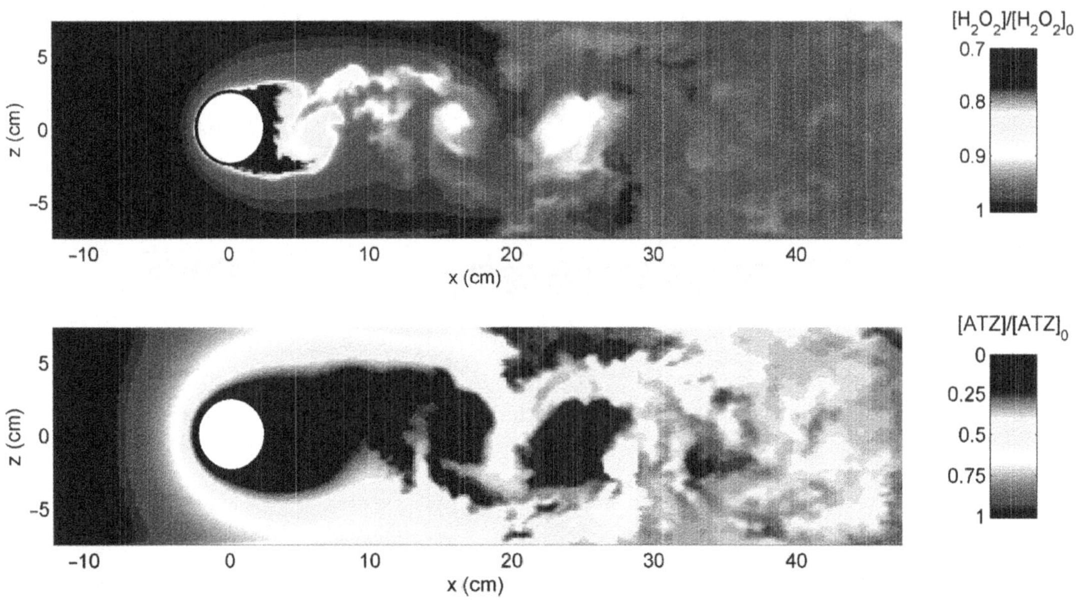

Figure 5.18 Snapshot of the reduction of hydrogen peroxide (upper panel) and the degradation of atrazine (lower panel) calculated by the LES model. A movie can be found at: http://dx.doi.org/10.4121/uuid:dfff9d45-75c4-455c-848a-d50bc4a09087

Besides the Eulerian approach, the degradation of atrazine was also computed in a Lagrangian framework from the dose distribution determined by the particle tracks (Chapter 2). Moreover, the atrazine degradation was calculated with and without the assumption that the reduction of peroxide can be neglected. The results of the average atrazine degradation for these approaches are shown in Table 5.4, showing that neglecting the reduction of peroxide had little effect (less than 1% difference). Also, the Lagrangian approach gave similar results as the Eulerian approach, showing that both methods could be used to calculate the degradation.

Table 5.4 Results of oxidation calculations for different modelling approaches.

Modelling approach	Degradation atrazine (%)
LES model, Eulerian, with peroxide reduction	65.4
LES model, Eulerian, peroxide constant	65.9
LES model, Lagrangian, with peroxide reduction	65.5
LES model, Lagrangian, peroxide constant	65.8
k-ε model, Lagrangian, with peroxide reduction	67.7
k-ε model, Lagrangian, peroxide constant	67.9

The degradation of atrazine under similar conditions as the example above was computed for the various cross-flow UV lamp geometries using the Lagrangian approach. The results for the k-ε model and the LES model are shown in Figure 5.19. The LES model predicted a lower removal of atrazine, even for the reference cylinder. The differences between the modelling approaches were again largest for the upstream baffle configuration, where the k-ε model predicted 15% more degradation of atrazine. For oxidation, the effect of placing baffles was comparable as for disinfection, although the differences were less pronounced. Placing the baffle downstream resulted in the highest increase in degradation, whereas placing a baffle upstream had the opposite effect.

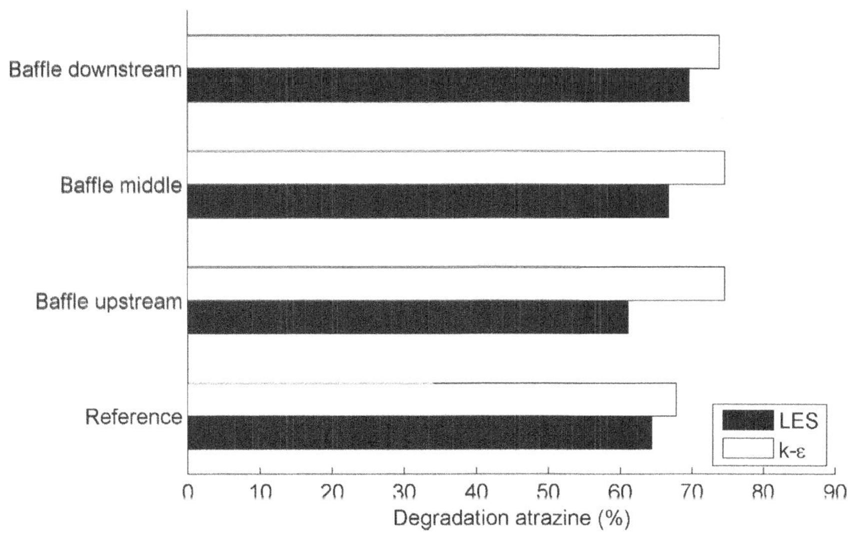

Figure 5.19 Overview of the degradation of atrazine, predicted by the LES model and k-ε model for the various geometries.

5.4.7 Discussion

The differences between the modelling approaches become more apparent by considering the movements of the individual particles (Figure 5.20). First of all, for the reference case, the positions of the particles at different time steps were considered for both models. From earlier observations we have seen that the

LES model showed smaller residence times than the k-ε model (§5.4.4), whereas the peak in the dose distribution was found at a similar dose. These observations can be related to the particles positions and the UV dose received by the particle. Clearly, the way the particles had moved was different for both models. The k-ε model predicted a more smooth concentration distribution of particles due to the smaller velocity gradients and small wake region. The region downstream of the lamp was therefore quickly filled with particles. From the particle positions calculated by the LES model, the turbulent flow features were better recognisable. For example, in the case of the reference cylinder at $t/T_h = 0.25$, the large wake was visible by the particles that moved around it. An instant later, the large eddies shed from the cylinder were visible: they caused mixing of low and high dose particles. Subsequently, particles were trapped inside the wake region, where they received very high doses. The LES model predicted the quickest moving particles due to the higher maximum velocities, which occurred at a small distance above as well as below the lamp. These particles did not necessarily have the smallest dose. In fact, these particles moved close to the lamps, where the UV irradiances were high. Instead, the particles with the lowest doses had moved along the upper and lower wall. After they had passed the lamp, they were – after a while – redistributed over the height due the increased turbulent mixing downstream of the lamp. Since the velocity fields near the outer walls were almost similar for both models, the critical particles, which are the particles with the lowest doses, were predicted the same. As a result, the disinfection prediction for the reference case was equal for both models.

Next, the particle positions for the baffle upstream configuration are considered in Figure 5.20 as well. For this configuration the differences in disinfection levels predicted by both models were large. The critical flow trajectories from the reference configuration - the particles with the lowest doses moving along the outer walls, were forced to move around the baffle. The differences between the models is most clearly observed at $t/T_h = 0.25$. The k-ε model predicted a spreading of particles over the height, including a large group of particles directly downstream of the lamp, whereas the LES model predicted particles close to the outer walls, which had moved along with the jets above and below the lamp. Due to the high velocities here, the UV doses remained small. Further downstream, these particles were redistributed over the height, but at such a distance from the lamp that the doses remained small. Therefore, at $t/T_h = 0.375$, the LES model predicted large numbers of particles close to the outflow, which were responsible for the low residence times and low doses. For the k-ε model, the particles had moved slower and closer to the lamps, so that the peak in the dose distribution was found at a higher dose. Consequently, the predicted disinfection levels differed largely. The longer tail in the RTD predicted by the LES model was caused by the large wake region, which is visible by the particles trapped inside this region.

5.5 CONCLUSIONS

Measurements were conducted to investigate hydraulic changes that reduce the wake size downstream of UV lamps. Dye visualization and LDA velocity measurements showed that the size of the wake was sensitive to local turbulent properties of the flow. Placing a grid upstream of the lamp resulted in a shorter wake length. Increasing the roughness of the lamp surface did not have the desired effect to advance the drag crisis to a lower Reynolds number, which should reduce the wake size. On the contrary, the wake size was increased considerably. Streamlining the lamp resulted in a disappearance of the wake in case the Reynolds number was high enough (above $Re = 24000$).

The LES model and k-ε model were evaluated for the flow around a single cross-flow UV lamp. The k-ε model largely under predicted the wake formation downstream of the reference lamp, whereas the LES model predicted the flow fields accurately. Using both modelling approaches, the sensitivity of the residence time distribution, dose distribution and disinfection to the local velocity field was investigated.

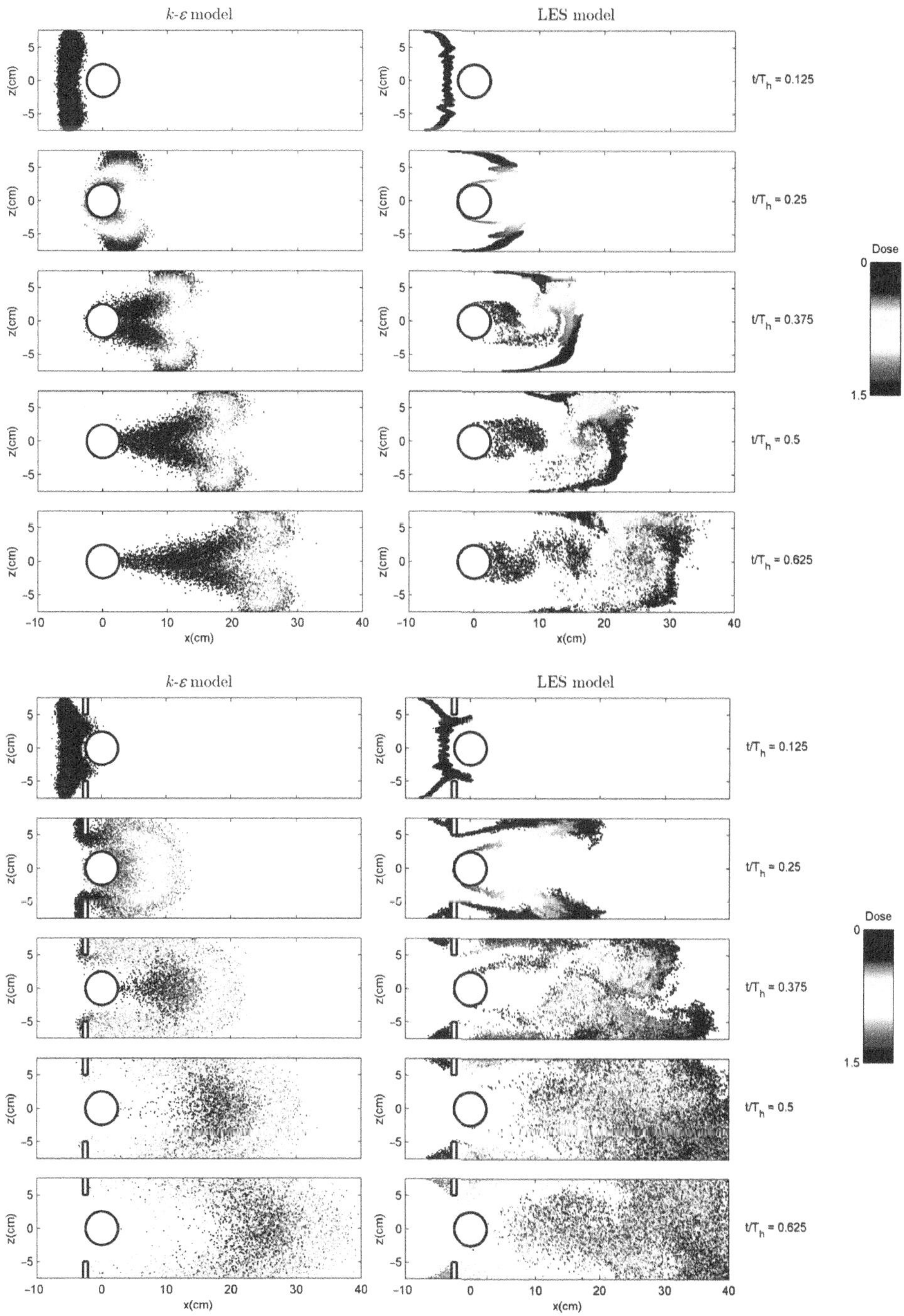

Figure 5.20 Particle positions calculated by the *k*-ε model and LES model at different time steps for the reference cylinder (upper panels) and upstream baffle (lower panels). The normalised dose is indicated by the colour of the particle. A movie can be found at: http://dx.doi.org/10.4121/uuid:63ef98c8-5a01-442a-9caf-80106576a171

The effect of placing baffles around a single UV lamp were demonstrated by the CFD model with a large-eddy simulation. By placing baffles the residence time distribution became wider with respect to the reference situation without baffles, however, the disinfection was in some cases increased. The disinfection efficacy was very sensitive to the position of the baffles. Placing the baffles a little upstream of the centre of the lamp (distance of $0.5D_{\mathrm{lamp}}$) resulted in a less efficient disinfection than for the case without a baffle. Moving the baffle a little downstream (up to a distance of $0.5D_{\mathrm{lamp}}$ downstream of the centre of the lamp) resulted in an increase in disinfection.

The weaknesses of RANS models, that do not resolve the turbulent motions, were demonstrated. By simplifying the turbulent flow inside the UV system, flow fields are predicted inaccurately in time and space. This type of modelling does not capture all the complex turbulent flow details, and the sensitivity of baffle positions on the local hydraulics may be addressed wrongly. The effect of this inaccurate representation of the hydraulics on the disinfection prediction depends on the local geometry. As shown here for the baffle configurations, the predictions of the k-ε model become inaccurate in situations where large velocity differences are present and large wakes can develop. The latter may not be the case for full-scale installations, where the wakes are more confined due the interaction with other lamps. This is explored in Chapter 6, where the model approaches are validated by experimental results for a bench-scale UV reactor. Nevertheless, the investigations on the single cross-flow UV lamp show that one must be aware of the limitations of the turbulence models. More advanced models, such as the LES model, result in a more precise disinfection prediction, but are computationally more expensive.

Chapter 6

UV systems*

Where Chapter 5 focuses on the details of a single UV lamp system, in this chapter bench-scale and full-scale UV systems are considered. We provide a validation of the CFD modelling of a bench-scale UV reactor by a combination of measurement techniques. Lastly, we investigate the design of new UV systems using the modelling tools described in Chapter 2.

6.1 INTRODUCTION

The experimental and numerical results of a single cross-flow UV lamp, described in Chapter 5, showed that the local (turbulent) flow features were very important for the predicted disinfection or oxidation. The RANS approach with a k-ε turbulence model may in some cases overestimate the disinfection efficacy significantly with respect to the more advanced LES model. The next step is to evaluate whether this overestimation also occurs in more realistic UV systems, and whether this can be confirmed by experiments. A large set of experiments were therefore conducted in a bench-scale UV reactor. The experiments include measurements of flow velocities, residence time distributions, dye visualisations, dose distributions, microbial inactivation and degradation of chemicals. This combination provides a thorough validation of the CFD modelling of the UV reactor, which is important to guarantee that the predicted performance by CFD modelling is accurate.

After validation of the CFD model, the model was used to optimise UV reactor performance. Several authors evaluated the disinfection performance of different reactors using CFD (Wright and Hargreaves, 2001; Sozzi and Taghipour, 2006; Pan and Orava, 2007). They found differences in the disinfection performance between the various reactors, which were related to the hydraulics of the systems. Modifications to the geometry of the reactor, such as baffle plates or rings, can result in an increase in disinfection capacity (Blatchley III et al., 1998; Chiu et al., 1999). Among the large number of reactor

* Parts of the research described in this chapter are published in: Wols, B.A., Shao, L., Uijttewaal, W.S.J., Hofman, J.A.M.H., Rietveld, L.C., and van Dijk, J.C. Evaluation of experimental techniques to validate numerical computations of the hydraulics inside a bench-scale UV reactor. Chemical Engineering Science 65(15): 4491–4502, http://dx.doi.org/10.1016/j.ces.2010.04.013, http://dx.doi.org/10.1016/j.ces.2010.04.013
Wols, B.A., Hofman, J.A.M.H., Beerendonk, E.F., Uijttewaal, W.S.J. and van Dijk, J.C. A systematic approach for the design of UV reactors using CFD. AIChE Journal 57(1): 193–207, http://dx.doi.org/10.1002/aic.12255, http://dx.doi.org/10.1002/aic.12255
Parts of the data can be accessed through http://dx.doi.org/10.4121/uuid:c1ac7344-1419-4398-ba13-c757551c303f

types, it is not known beforehand which reactor type has the best performance with respect to disinfection or oxidation, and if such a reactor is the best possible reactor. A systematic approach for the design of UV reactors is therefore followed that makes use of CFD modelling. Design aspects such as placing mirrors, determining the lamp size and placing reactors in series are considered. The various designs are subjected to the following criteria: disinfection performance, oxidation performance and energy consumption. The potential in energy reduction for existing UV reactors and new reactor designs is explored as well.

The experimental results for the bench-scale UV reactor are presented in §6.2, whereas the CFD modelling of this reactor by both modelling approaches (the k-ε model and LES model) are presented in §6.3. In §6.4, the optimal design of a UV system is studied by assessing a large number of UV reactors.

6.2 EXPERIMENTS OF A BENCH-SCALE UV REACTOR

6.2.1 Setup

A bench-scale UV reactor was constructed around four UV lamps placed perpendicular to the flow direction (see Figure 6.1). Upstream of the reactor, a pipe ($D = 0.125$ m) with a length 50 times the pipe's diameter was used to obtain a fully developed turbulent velocity profile at the entrance of the UV reactor. The outer geometry of the reactor itself consisted of a cylinder ($D = 0.3$ m) with a depth of 0.15 m, placed perpendicular to the direction of the inflow and outflow pipes. Inside the reactor, four cylindrical quartz tubes were placed. The outflow pipe had the same cross-sectional area as the inflow pipe. Flow rates were varied between 1 m^3/h and 5 m^3/h, which resulted in a mean velocity in the pipe ranging from 0.023 m/s ($Re = 2.8 \cdot 10^3$) to 0.11 m/s ($Re = 14 \cdot 10^3$), indicating that the flow was turbulent. Two measurements series were conducted: a series to measure the hydraulics and a series to measure dose distributions, microbial inactivation and chemical degradation.

Figure 6.1 Photograph of the UV reactor (upper panel), where four lamps are placed perpendicular to the flow direction; the geometry of the UV reactor used in the CFD model is shown in the lower panel.

For the first measurements series, no UV lamps were installed in the quartz tubes. The residence time distributions (RTD) were obtained by injecting a tracer of salt (sodium chloride) water and measuring the electrical conductivity of the water upstream and downstream of the reactor. The injection of salt water occurred at a large distance upstream of the reactor (more than 50 times the diameter of the pipe) to ensure a good mixing of the salt water over the cross-section. The salt concentration was measured in the middle of the pipe, 0.45 m upstream as well as 0.45 m downstream of the centre of the reactor. The salt plume had an increased salinity of 0.1 g/L, which was low enough to prevent the occurrence of density currents and high enough to be detected accurately.

Assuming that residence time distributions can be described by an open dispersion model (Levenspiel and Smith, 1957; Aris, 1959; Martin, 2000), Péclet numbers were obtained from the difference in the variance of the input and output RTD. The injection of a passive tracer was used to visualise the flow fields. Dye (potassium permanganate, $KMnO_4$) was injected upstream of the reactor. Because of the dye, the flow was visible through the transparent covers mounted on the flanges of the reactor. The dye, injected 1 m upstream of the reactor, allowed for mixing over the entire cross-section upstream of the reactor. The dye slowly filled the reactor and the injection was stopped when the water in the reactor turned completely purple. Fresh water entered the reactor again and the dye slowly drained out. A movie camera (3 MP, 25 Hz) recorded the spreading of the dye through the reactor. The flow fields were measured by means of laser Doppler anemometry (LDA), explained in §5.2.1.

In addition to the hydraulic measurements, biodosimetry and Lagrangian actinometry experiments (Blatchley III et al., 2008) were performed in the same UV reactor. The transparent flanges were replaced by stainless steel flanges and four monochromatic UV lamps were placed in the middle of the quartz tubes. For the biodosimetry experiments, MS2 phages and *Bacillus subtilis* spores were used as challenge organisms. The MS2 phages were prepared in the microbiology laboratory of KWR, where the phages were cultivated using the cells of a host bacteria. The *Bacillus subtilis* spores were obtained from Vienna (ÖNORM, 2001) using the 2007 batch. The dose distributions were determined by Lagrangian actinometry, where fluorescent microspheres were added to the water. The dyed microspheres were prepared at Purdue University by conjugating a photosensitive substance S (which represents (E)-5-[2-(methoxycarbonyl)ethenyl]cytidine) to polystyrene microspheres through a streptavidin-biotin linkage (Blatchley III et al., 2006).

The experimental set-up is shown in Figure 6.2. Nieuwegein tap water is entering the system after a 1 μm filter unit to remove coarse particles. The transmittance was reduced by adding a PHBA solution (4-hydroxybenzoic acid 99%, from Sigma Aldrich, Cas. Nr. 99-96-7.), which absorbed light photons (molar extinction at 246 nm is 11500 L/mol/cm, Modi (1994)). During the experiments, two samples at the influent and two samples at the effluent were taken to determine the transmittance by spectrophotometry. A UV-vis spectrophotometer (Unicam UV530, calibrated in June 2009) measured the absorbance of the water sample in a 1 cm cuvette. Filtration of the samples before analysis by the spectrophotometer was not necessary, indicating that the effect of particle scattering could be neglected. The transmittance for each experimental condition was determined by averaging the values of the influent and effluent samples. A stock solution containing 5 L Nieuwegein water (filtered by a 1 μm membrane), MS2 phages, *Bacillus subtilis* spores and dyed microspheres (DMS) was prepared for each experimental condition. This solution was added to the system by means of a peristaltic pump (flow rate of 10.1 L/h). After the addition of PHBA and the stock solution, a static mixer ensured a good mixing with the main stream. Downstream of the static mixer an influent sample port was placed. An effluent sample port was placed after a static mixer downstream of the reactor. For the biodosimetry experiment, five samples at the influent and five at the effluent sample port were taken to determine the microbial inactivation.

For the analysis of the concentration of micro-organisms, the *Bacillus subtilis* samples were pasteurized at 60 °C during 30 minutes, after which they were diluted and put on plates with bouillon agar (in duplicate).

Incubation of the plates occurred at 36 °C during 24 hours, after which the spores were counted. The MS2 phage samples were analysed in accordance with ISO 10705-1 (1995). These samples were diluted and put on plates with host strains WG49 and bouillon agar (in duplicate). After incubation of 18 hours at 37 °C, the colony-forming units were counted.

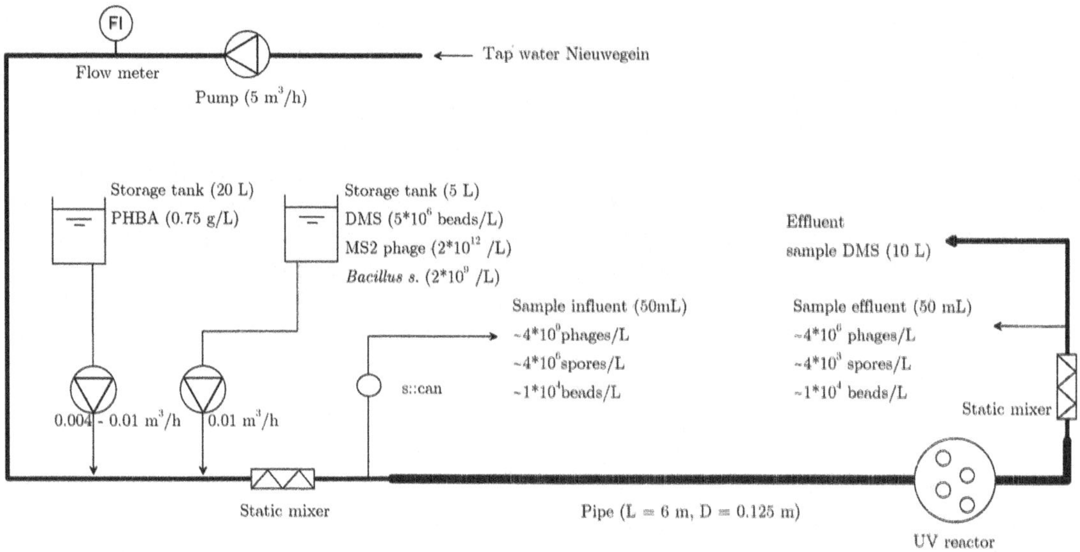

Figure 6.2 Experimental set-up of the UV system as used for the biodosimetry and Lagrangian actinometry experiments.

For the Lagrangian actinometry, a sample of the stock solution was taken as a reference for the microspheres. Three 10 L effluent samples were collected directly from the effluent of the system, containing around $1 \cdot 10^5$ microspheres. The microspheres were separated by means of membrane filtration (3 μm), followed by resuspension in an aqueous buffer, and centrifugation. These samples were wrapped with Al foil and were packed on ice, after which they were shipped to Purdue University for analysis by flow cytometry. More details on the Lagrangian actinometry measurement technique are provided by Blatchley III et al. (2006, 2008).

Two different lamps (Table 6.1) and two different transmittances (through 1 cm) were used, resulting in three different experimental conditions (Table 6.2). A UV radiometer (SUV20.2A2Y1R/150/UVD6, Il Metronic Sensor Technik GmbH) was used to determine the UV irradiance at a single position in the reactor (located at $x = 8.5$ cm, $y = 8.5$ cm from the middle of the reactor, which is 3 cm away from the outer wall). The irradiance measurements were used to confirm that the irradiated UV remained stable during the experiment. However, the absolute values of the UV sensor were not calibrated.

Table 6.1 Properties of the two types of UV lamps used in the bench-scale experiments.

Lamp type	Power UVC (W)	Arc length (m)	Diameter (mm)
Hereaus NNI125_84XL	42	0.79	15
Van Remmen PUV300_300W	115	1.47	19

Table 6.2 Experimental conditions for the biodosimetry and Lagrangian actinometry experiments. The experiments are numbered from a high expected mean dose to a low one.

Exp #	P_{lamp} (W)*	$P_{reactor}$ (W)**	T_W (%)	Water temp (C)	Air temp (C)†
1	42	31.9	87.7	12.6	22
2	115	46.9	78.4	12.8	36
3	42	31.9	78.5	12.6	22

*The UVC power of each lamp over the length of the lamp.
**Total UVC power emitted by the four lamps in the reactor (over the reactor width of 15 cm).
†Air temperature of the box in which the reactor and UV lamps were installed.

Next to the bench-scale experiments, collimated beam experiments were conducted to determine the dose-response curves of the micro-organisms and to determine the fluorescence excitation of the microspheres under different doses. Everyday of bench-scale testing, a Petri dish containing 60 mL of Nieuwegein water (filtered by a 1 μm membrane) and microspheres ($1 \cdot 10^8$ beads/L) was irradiated in the collimated beam apparatus. The microspheres were subjected to doses ranging from 0 to 180 mJ/cm^2 with steps of 20 mJ/cm^2. The highest dose corresponded to three times the expected mean dose in the UV reactor. After each UV dose increment a 1 mL sample was taken from the Petri dish. This sample was used to determine the fluorescence excitation of the microspheres for that particular dose. For the collimated beam test of the micro-organisms, *Bacillus subtilis* spores ($5 \cdot 10^7$ spores/mL) and MS2 phages ($1 \cdot 10^9$ phages/mL) were added to a 50 mL Petri dish containing Nieuwegein water (filtered by a 1 μm membrane), and subjected to UV doses of 0, 20, 40, 60, 80 and 100 mJ/cm^2. A low pressure lamp (PLL90W) was used to generate the required UV doses. A SED 240W sensor (International Light) with an IL 1700 Model Research Radiometer measures the UV radiation, whereas the absorbance of the sample was measured by spectrophotometry. Based on these values and the set-up of the system (volume of the Petri dish, distance from the sample to the lamp), the exposure times were determined from an Excel-sheet developed by Bolton and Linden (2003).

In the Lagrangian actinometry measurements, the fluorescence intensity (FI) of a microsphere subjected to UV was measured by flow cytometry. The microspheres were discriminated against ambient particles by a technique called gating, which is often used in flow cytometry to identify the particles of interest by the known optical characteristics (size and FI) of these particles (Blatchley III et al., 2006). The collimated beam tests provided the fluorescence response of the microspheres under different doses. Due to small variations in the characteristics of the microspheres (size, shape and surface loading of the dye) and errors in FI measurements by flow cytometry, a distribution of FI was measured for each UV dose established in the collimated beam. Subsequently, the fluorescence intensity of each microsphere in the sample collected from the bench-scale reactor was determined by flow cytometry. A wide distribution of FI is obtained due to the differences in UV dose received by the microspheres and the above mentioned variability in the FI measurements. The UV dose distribution of each bench-scale reactor sample could then be determined by using a numerical scheme (deconvolution) that couples the FI distribution of this sample to the FI distributions of the collimated beam samples (Blatchley III et al., 2006). The flow cytometry analyses of the samples and the deconvolution of the data were performed by people from Purdue University.

6.2.2 Residence time distributions

In Figure 6.3, the residence time distributions obtained from the electrical conductivity measurements are plotted. The input signal showed a spreading in residence times as a result of the dispersion in the pipe circuit upstream. After the salt plume had traversed through the reactor, the residence time distribution became wider. The mean residence times were in the range of the theoretical hydraulic residence time, except for the lowest flow rate. Here, the mean residence time was found to be two or three times larger than the hydraulic residence time. Also, the signal contained more noise than the signals at the higher flow rates. Reynolds numbers in the pipe were around 2500 for the lowest flow rate, so the flow was close to the laminar regime, resulting in a small degree of mixing, which explains the small peaks in the signal. Since the flow was forced to move around the lamp, concentration differences over the cross-section were enforced. Due to a small lateral mixing, these concentration differences extended over a large distance and caused these noisy concentration profiles. Accordingly, Péclet numbers were small, confirming the low degree of lateral mixing. Therefore, the samples from the middle of the pipe could not be representative for the total cross-section, which explains that the measured hydraulic residence times deviated from the theoretical ones at the lowest flow rate.

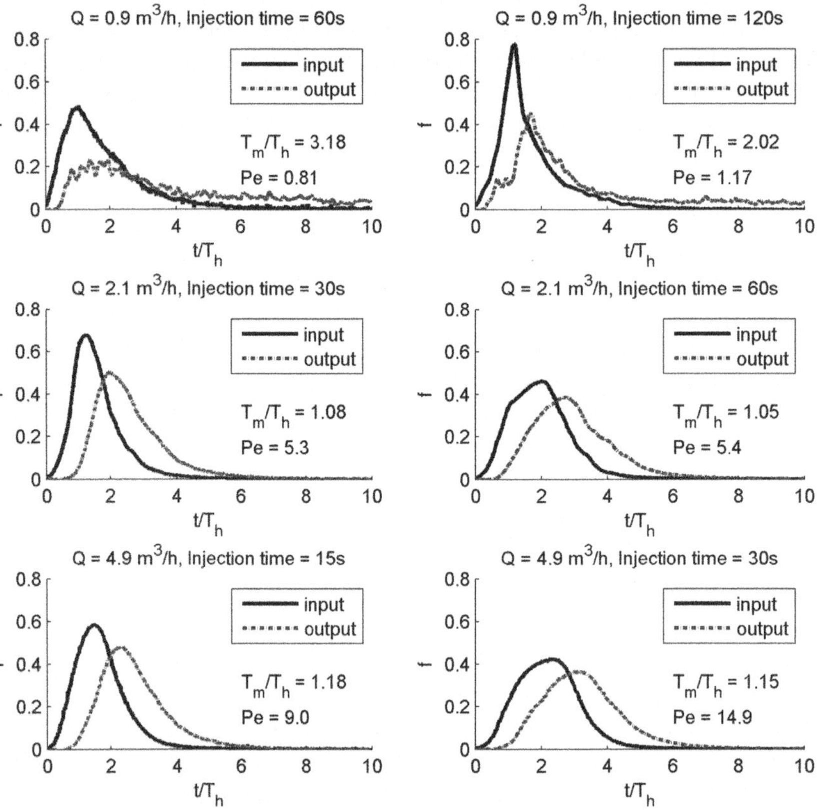

Figure 6.3 Residence time distributions, obtained from the salt concentration measurements, are shown for the sample port upstream (input) and sample port downstream (output) of the UV reactor. The time is normalised with the hydraulic residence time (T_h).

At the higher flow rate of 2.1 m^3/h, the Péclet numbers increased by a factor of five, mainly due to the transition from nearly laminar flow to fully turbulent flow. As a result, the concentrations were more uniformly spread over the cross-section, reducing the noise in the RTD. The increase in the Péclet number from a flow rate of 2.1 m^3/h to 4.9 m^3/h can be explained by the increase in advective transport, whereas the diffusive transport remained the same.

6.2.3 Dye measurements

In Figure 6.4, the concentration of dye averaged over the reactor is plotted against time, and some frames of the dye movie corresponding to different stages of the filling and emptying process are shown. Visual observations gave the remarkable impression that emptying the reactor filled with dye seemed to take much longer than filling the reactor with dye. An explanation might be that differences between low concentrations of dye and fresh water were clearly visible, whereas differences between the maximum concentration of dye and some smaller concentration were hardly visible. Moreover, the videos were a 2D representation of a 3D situation, so the concentration differences in a lateral direction could not be observed. During the filling process, a high dye concentration in the middle and a lower dye concentration at the sides might give the impression that the reactor was completely filled. Yet, at the same stage during the emptying process, no dye concentration in the middle and a low dye concentration at the sides showed that the reactor was not yet empty. In the following paragraphs, we focus on the emptying of the reactor, since it most clearly visualises the recirculation zones in the reactor.

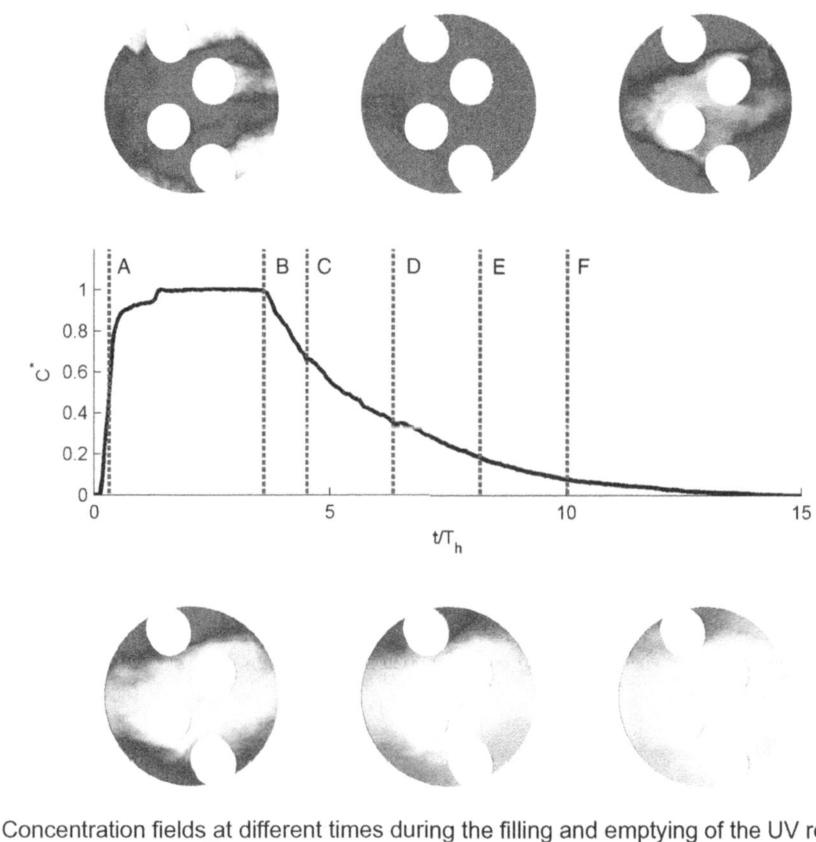

Figure 6.4 Concentration fields at different times during the filling and emptying of the UV reactor with purple dye for a flow rate of 5 m^3/h. The measured concentrations over the cross-section, collected from the video recordings, are shown at 6 different time levels (A - F). The flow is coming from the left. In the middle panel, the normalised concentration (C^*) averaged over the cross-section is shown as a function of dimensionless time (normalised with the hydraulic residence time). A movie can be found at: http://dx.doi.org/10.4121/uuid:06d6b6e3-3fe1-4502-825b-6627a5deaa5f.

Three regions inside the system were considered: the complete reactor, a flow region and a dead zone region. The flow region was regarded as the horizontal region in the middle of the reactor, with a height equal to the pipe diameter. The dead zone region was the region above and below the flow region. In Figure 6.5, the mean concentrations at these three regions are shown for different flow rates. The concentrations were normalised with respect to the maximum concentration of each region. The concentration in the flow region started to diminish linearly, caused by the advection of the flow region, and ended with an exponential tail (due to the slowly emptying of the dead zone). The concentration in the dead zone region started to diminish linearly at a small rate and after a certain time (typically the time scale needed for the emptying of the flow region) the diminishing became exponential, causing a long tail in the concentration time curve. The exchange of concentration between the flow region and the recirculation zone is a complex process depending on the local turbulent properties: turbulent eddies mix highly concentrated fluid with less concentrated fluid from the main flow.

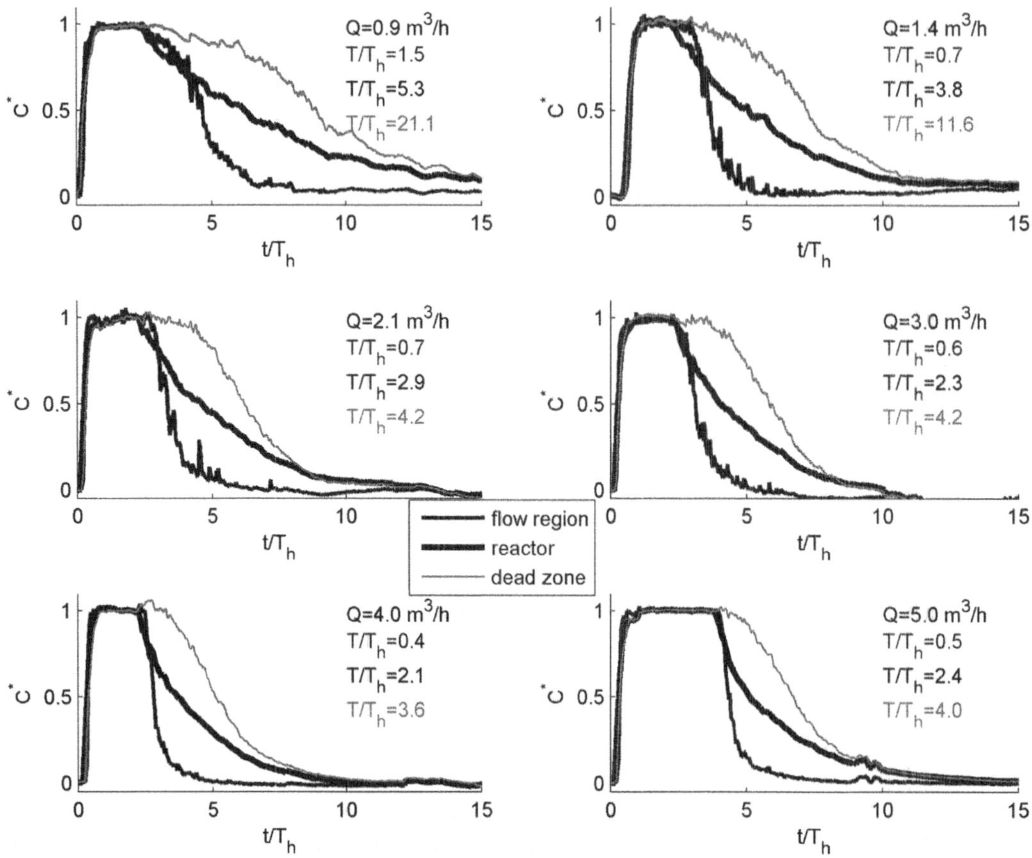

Figure 6.5 The measured dye concentrations, averaged over different regions inside the UV reactor (flow region at the middle of the reactor, dead zone region at the upper and lower part of the reactor, and the complete reactor region), are shown as a function of dimensionless time for the different flow rates.

Characteristic time scales were estimated for each region, as shown in Figure 6.5. The time scale for emptying the flow region was typically half that of the hydraulic residence time, whereas the dead zone region took at least four times the residence time to empty. In general, the time scales lessened at higher flow rates. For the lowest flow rates, these values were significantly higher, mainly caused by the small degree of lateral mixing inside the reactor (due to the nearly laminar flow). Also, the signals were smoother for the higher flow rates because the concentration differences were lessened due to a higher lateral mixing. The time scale required for emptying the entire reactor varied from two times the hydraulic residence time at the highest flow rates up to a factor of five at the lowest flow rates.

6.2.4 Flow fields

Velocities were measured over different horizontal and vertical lines at the middle plane of the reactor. These lines were chosen at typical positions inside the reactor: upstream of the lamps to characterise the inflow conditions and downstream of the lamps to investigate the effect of the lamps on the flow field. At the highest flow rate ($Q = 5\,m^3/h$), these lines were also measured at five different planes over the width of the reactor to examine the symmetry. The velocity field at the middle plane, derived from the mean velocities at the measured positions, is given in Figure 6.6 (for flow rates of $2.1\,m^3/h$ and $5\,m^3/h$). Contraction of the flow between the lamps is visible, resulting in small residence times (short-circuiting). The flow was separated in the wake of the lamps, causing recirculation zones that induced large residence times. Near the bottom of the reactor, an upward flow is visible for both flow rates, indicating that a large recirculation was present here. The difference in Reynolds number at the inflow ($Re = 6000$, and $Re = 14\,000$, respectively) caused small differences in the flow patterns. For example, the wake behind the upper right lamp was larger for the higher Reynolds number.

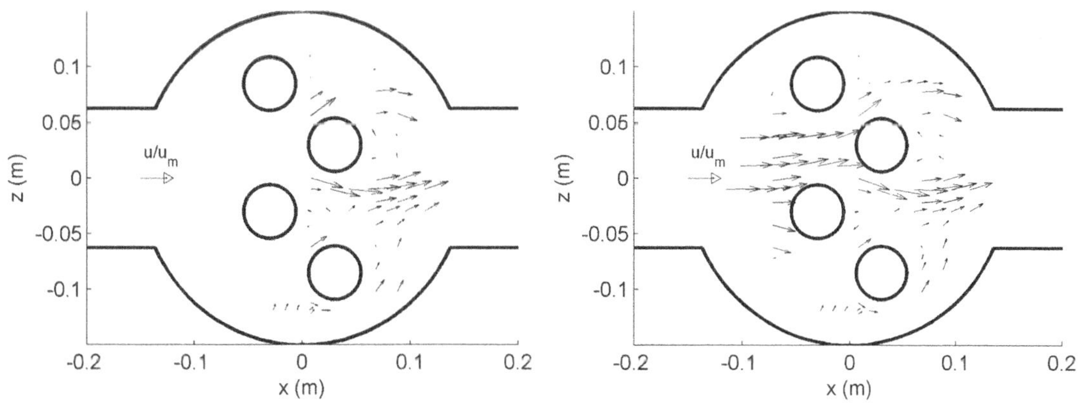

Figure 6.6 Velocity vectors, mean velocities measured with the LDA device for a flow rate of $2.1\,m^3/h$ (left panel) and $5\,m^3/h$ (right panel).

In Figure 6.7, the streamwise velocities at different planes over the width are given at the middle plane of the reactor and at planes at $\frac{1}{4}$ and $\frac{3}{4}$ of the reactor width (the dotted lines represent the measured positions). Upstream of the lamps, the velocities at the outer planes were lower than at the middle plane due to an abrupt lateral expansion from the pipe to the reactor. This effect was reduced downstream of the lamps. Differences between the centre and the sides were only observed at the mixing layer between the recirculation zone of the lamp and the main flow. Concerning the symmetry, small differences between the velocities at both sides of the reactor were observed. However, minimal deviations from the symmetrical geometry, for example the orientation of the reactor or orientation of individual lamps, may already induce a large asymmetrical flow.

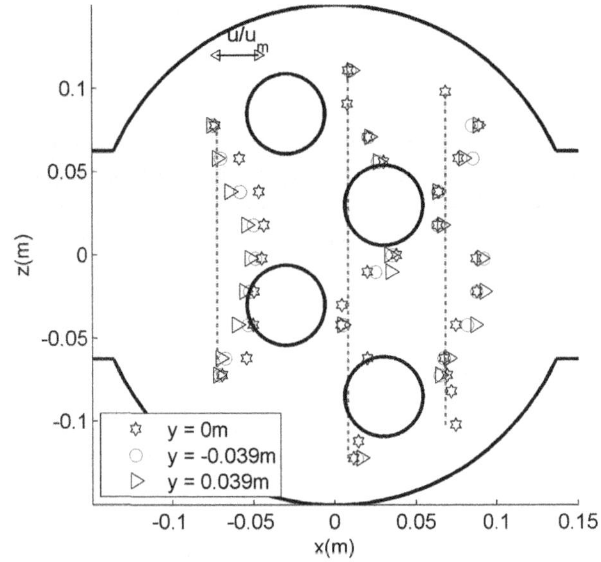

Figure 6.7 Variations in streamwise velocities over the width of the reactor. The dotted line corresponds to the line at which the measurement was taken.

The Reynolds stresses were considered as a means to investigate the mixing behaviour. These stresses represent the fluctuating part of the velocities caused by the turbulence and unsteady behaviour of the flow, which contributes to the mixing. For the highest flow rate (5 m^3/h), the three components of the Reynolds stress tensor (in 2D, q_{uu}, q_{uw}, q_{ww}, N/m^2) are plotted in Figure 6.8. The stresses are plotted along four vertical lines inside the reactor: upstream of the lamps, in between the lamps, and at two lines downstream of the lamps. In the middle of Figure 6.8, a 3D visualisation of the Reynolds normal stresses (in a streamwise direction) is plotted, showing the regions where the highest stresses occur. Upstream of the lamps, the stresses were relatively small; close to the lamps, a small increase is visible. Downstream of the lamps, the Reynolds stresses were much higher; the velocity fluctuations were increased due to the high shear and vortex shedding, resulting in a higher mixing. The Reynolds normal stresses in horizontal and vertical directions (q_{uu} and q_{ww}) were of the same order of magnitude, indicating that mixing in horizontal and vertical directions had the same intensity. In the wake of the lamps, the stresses were smaller, but increased rapidly towards the shear layer between the wake and the main flow. A similar pattern is visible for the Reynolds shear stresses (q_{uw}), which showed the highest values at the shear layer, where the exchange of momentum by the turbulent eddies was also highest. The Reynolds stresses at the bottom of the reactor were smaller than at the top, probably because the downstream part of the lower lamp was more confined by the reactor wall, preventing the development of large eddies.

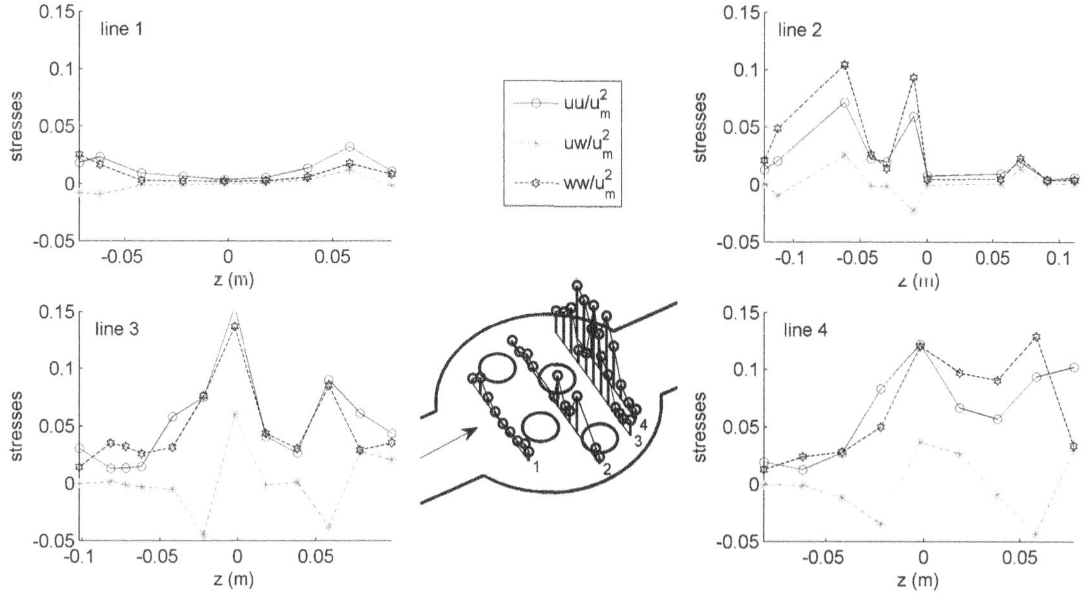

Figure 6.8 Reynolds stresses measured at different lines at the middle plane of the reactor. In the middle, the spatial distribution of the Reynolds stress q_{uu} is plotted.

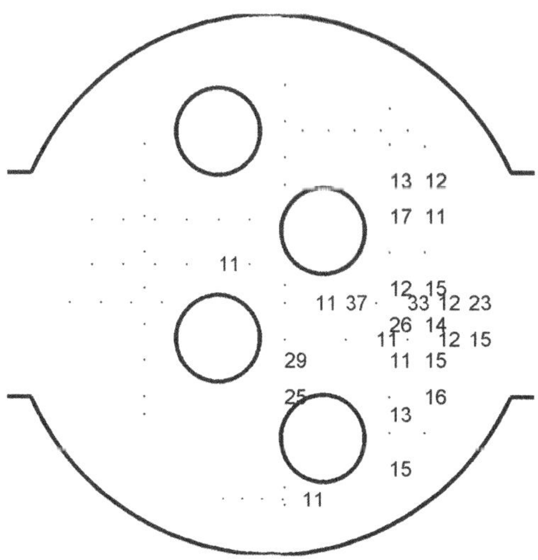

Figure 6.9 Strouhal numbers ($St \cdot 10^2$) derived from the instantaneous velocity measurements.

The Strouhal numbers were plotted at different locations in the reactor (Figure 6.9). Strouhal numbers less than 0.1 are not plotted because they are not related to vortex shedding (shown as the small dots). Because the peak in the energy spectrum at the vortex shedding frequency did not differ much from the energies at different frequencies, the Strouhal numbers calculated could be inaccurate, causing them to range from 0.11 to 0.37. But, there was a distinct pattern visible where frequencies related to vortex shedding occurred. These vortices were created at the downstream part of the cylinder and were detached periodically from the bottom and top of the cylinder. Downstream of these positions, Strouhal numbers of around 0.15 were observed, a little lower than the values measured at a single cylinder (Chapter 5). Higher frequencies occurred at the downstream side of the two cylinders to the right (especially at the vertical position in between the lamps). This might be explained by interference of the vortices originating from the different lamps. The occurrence of the vortices underlines the dynamic behaviour of the flow and complicates the CFD modelling.

6.2.5 Lagrangian actinometry

The dose distributions derived from the fluorescence response of the dyed microspheres are shown for the different experimental conditions in the bench-scale UV reactor (Figure 6.10). For each experimental condition three samples were taken: the dotted lines represent the dose distribution of each sample, whereas the bold line represents the average of the three samples. A large peak in the dose distribution (roughly between 20 and 60 mJ/cm^2) was distinguished that was caused by the microspheres that moved in conjunction with the main flow. This peak of low doses is most important for the performance of UV reactors. At doses above this peak (around 100 mJ/cm^2 and higher), the distribution was more noisy, because a smaller amount of microspheres received these higher doses. At the highest measured dose (180 mJ/cm^2), also a peak is visible which represents microspheres that were trapped in the large recirculations and received higher doses. These high doses could not be determined, because the maximum dose applied in the collimated beam was 180 mJ/cm^2. Inaccuracies might arise due to the interpolation and deconvolution of the FI distribution of each sample using the limited amount of doses from the collimated beam tests. For example, the peak at a zero dose is clearly an anomaly that was caused by the numerical procedure, since the fluorescence excitation of these microspheres showed a higher fluorescence excitation than the microspheres from the influent sample (that received zero dose).

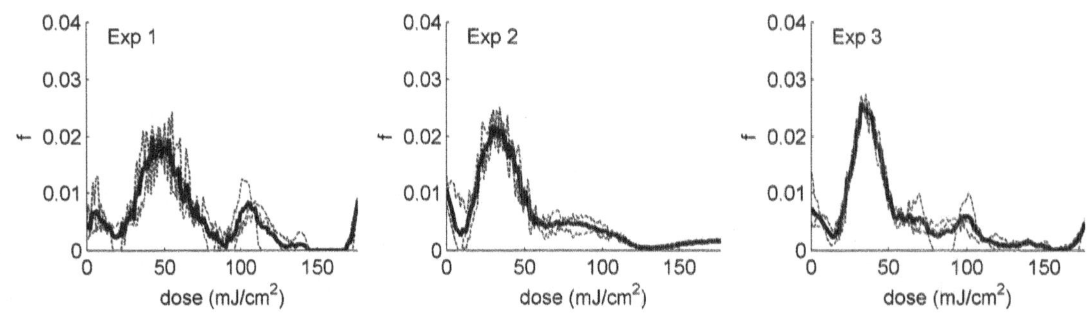

Figure 6.10 Dose distributions measured by the Lagrangian actinometry measurements. Three samples were taken (dotted lines), from which the averaged values were taken (solid line).

Due to these inaccuracies at the smallest doses and the limited range at the higher doses, the median dose D_{50} seemed to be the best parameter to characterises the dose distribution. The median dose represents the large peak in the dose distribution. In Table 6.3, the median doses calculated from the measured dose distribution are shown together with the lamp power, measured irradiation, theoretical scale factor for the dose (D_{scale}, §2.7.1) and measured microbial inactivation. The results are displayed from high expected UV doses (upper row) to low expected UV doses (lower row). This order was consistent for D_{scale}, the measured irradiation, and microbial inactivation of MS2 phages. The median doses determined from the Lagrangian actinometry measurements were not consistent with this order: the median dose in experiment 2 seemed to be low, whereas the median dose in experiment 3 seemed to be high. This illustrates the variability in the measured dose distribution by Lagrangian actinometry.

Table 6.3 Results of the Lagrangian actinometry (presented by the D_{50}) and biodosimetry experiments.

Exp†	$P_{reactor}$ (W)	T_W (%)	I (W/m^2)	D_{scale} (mJ/cm^2)	D_{50} (mJ/cm^2)	Microbial inactivation*	
						Bacillus	MS2
1	31.9	87.7	72 (±7)	128	53 (±3)	2.65 (±0.15)	3.13 (±0.14)
2	46.9	78.4	54 (±5)	102	40 (±3)	2.96 (±0.09)	2.71 (±0.08)
3	31.9	78.5	43 (±4)	70	41 (±2)	2.20 (±0.11)	2.34 (±0.03)

†Flow rate: 5 m^3/h, ± shows the standard deviation of different samples.
*Log units.

6.2.6 Disinfection

Results of the collimated beam tests

The collimated beam experiment for the micro-organisms, performed at the same period of the bench-scale experiment, had failed, because the concentration of *Bacillus subtilis* spores ($5 \cdot 10^7$ spores/mL) in the water sample was too high, resulting in a 1 cm transmittance of 14%. Due to the high concentration of *Bacillus subtilis* spores, the solution was highly turbid, resulting in a significant amount of scattering of UV radiation by the spores. Therefore, the Beer-Lambert law could not be used and the absorption could not be measured accurately. Consequently, due to the wrongly measured absorbance the irradiation times of the collimated beam samples were too high, resulting in too high log inactivation levels, both for *Bacillus subtilis* spores and MS2 phages.

Collimated beam experiments performed earlier (KWR (2004) and KWR (2006–2007)) were therefore used to investigate the dose-response behaviour of the micro-organisms (Figure 6.11). In these experiments the transmittances were higher (75%). For both challenge organisms, two different measurements are displayed, which were conducted under similar conditions for two identical batches of spores or phages. The results were in agreement with the dose-response curves from literature for both MS2 phages as *Bacillus subtilis* spores. A variation in the log inactivation of the micro-organisms is observed, in particular for the *Bacillus subtilis* spores. The following dose-response curve for *Bacillus subtilis* spores was used (ÖNORM, 2001):

$$\frac{N}{N_0} = 1 - \left(1 - 10^{-kD}\right)^{10^d}, \tag{6.1}$$

where D represents the dose (J/m^2). The constants $k = 0.0063$ and $d = 0.49$ were calibrated and confirmed by collimated beam test in November 2008 and August 2009 by the University of Vienna for the same batch of *Bacillus subtilis* spores as used in this experiment. For MS2 phages, a delayed Chick-Watson model was used to describe the dose-response behaviour using the kinetic constants from Table 2.1.

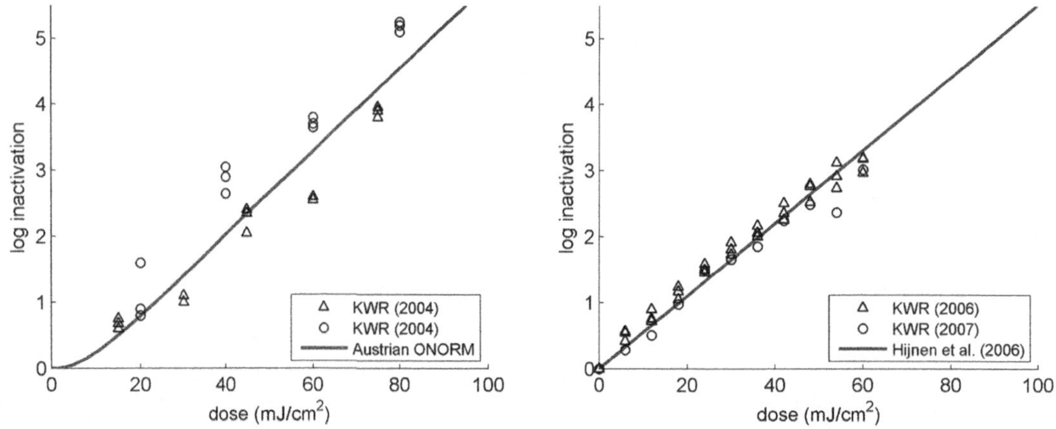

Figure 6.11 Collimated beam results for the inactivation of *Bacillus subtilis* spores (left panel) and MS2 phages (right panel) as well as the dose-response curves from literature: ÖNORM (2001) for *Bacillus subtilis* spores, and Hijnen et al. (2006) for MS2 phages.

Results of the bench-scale UV reactor tests

The microbial inactivations are shown in Table 6.3. The difference in concentration of micro-organisms between each effluent sample and the average concentration of the five influent samples was calculated. From these concentration differences, the average and standard deviation of the inactivation was calculated for each experimental condition. The standard deviations were small indicating that the microbial inactivations were accurately measured (deviation of around $\pm 5\%$). The measured inactivation of the MS2 phages followed the rank of expected UV doses (from high in experiment 1 to low in experiment 3), which was more consistent than the inactivation of the *Bacillus subtilis* spores. This observation is in agreement with the collimated beam tests, showing a larger variation in inactivation levels for the *Bacillus subtilis* spores than for the MS2 phages.

Decreasing the transmittance (from 87.7% in experiment 1 to 78.5% in experiment 2) resulted in a large decrease of the measured irradiance (40%) and of the characteristic dose D_{scale} (45%), however the reduction of the microbial inactivation (17% and 25%) and of the median dose (23%)was smaller. Increasing the lamp power (from 31.9 W in experiment 3 to 46.9 W in experiment 2) resulted in an increase in measured irradiance (25%) and microbial inactivation (35% and 16%), which was less pronounced than the increase in UVC power and D_{scale} (45%). Remarkably, varying the lamp power and the transmittance did not give such pronounced differences in dose distribution (D_{50}) and microbial inactivation as would be expected from the measured irradiance and D_{scale}. However, we must bear in mind that the irradiance measurement at a single point (aimed at a single lamp) does not characterise the radiation distribution in the complete reactor. Moreover, the D_{scale} is derived for an ideal system, which does not account for the local geometry (radiation losses near the walls, hydraulics etc.).

6.2.7 Oxidation

Results of the collimated beam tests

The unknown parameter in the modelling of oxidation processes is the factor $\sum k_i[N_i]$ that contains reactions of background substances with $OH\bullet$ radicals, see §2.5.2. The index *i* represents all the background substances in the water, except atrazine and peroxide. This parameter needs to be fitted from collimated beam experiments for each particular type of water used. We considered the degradation of atrazine, from which the kinetic constants were taken from literature, given in Table 2.2. In the case of demineralised water no background substances are present in the water, so $\sum k_i[N_i] = 0$. In Figure 6.12 the degradation of atrazine in demineralised water and Dunea water is shown (KWR, 2009). For demineralised water the predicted degradation by the model closely matched the measured degradation, confirming that the kinetic constants taken from literature were accurate. For Dunea water we fitted a factor $\sum k_i[N_i] = 5 \cdot 10^4\ s^{-1}$, which agreed well with the measured data for all peroxide concentrations. This factor is also used in the CFD model to calculate the degradation in the bench-scale UV reactor.

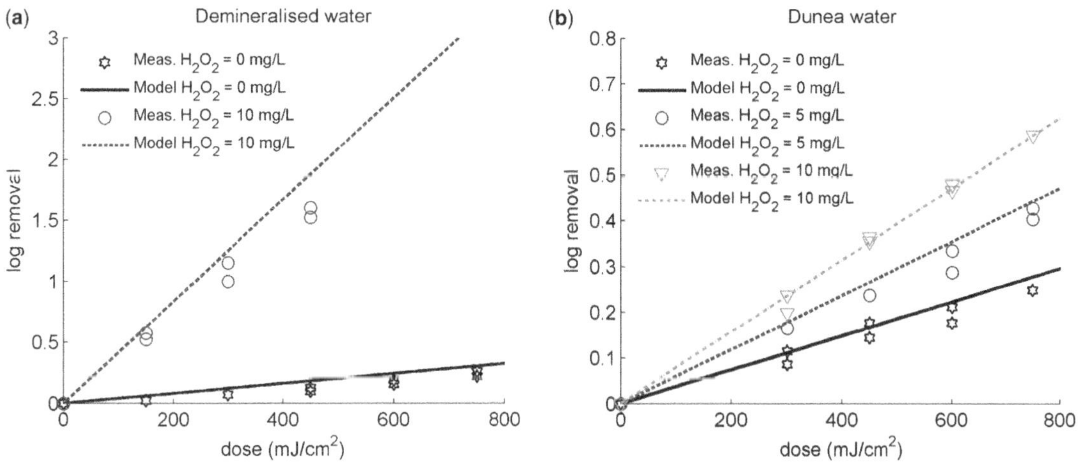

Figure 6.12 The collimated beam results are compared to the oxidation model for demineralised water (left panel) and Dunea water (right panel), KWR (2009).

Results of the bench-scale UV reactor tests

The degradation of atrazine was measured in the bench-scale UV reactor for different flow rates and different concentrations of peroxide, Table 6.4. Without peroxide, the degradation of atrazine by photolysis was relatively small. By adding peroxide, the degradation of atrazine was roughly doubled. Increasing the flow rate resulted in a lower degradation; the degradation is inversely proportional to the flow rate, which is in accordance with eq. 2.31, where the mean dose is inversely proportional to the flow rate.

Table 6.4 Results of the degradation of atrazine measured in the bench-scale reactor using Dunea water.

Flow rate (m^3/h)	Peroxide (mg/L)	Atrazine removal (%)†	Peroxide (mg/L)	Atrazine removal (%)†
2.1	0	21.4	10	42.9
3.0	0	14.3	9.4	29.3

†Total lamp power UVC: 88 W, transmittance water: 78%.

6.2.8 Discussion on the experimental results

The large set of measurements provided an extensive overview of the impact of the hydraulics in the bench-scale reactor. Visualisation studies with dye showed that recirculation zones occurred in the upper and lower part of the reactor, where the dye slowly diminished. Also, the vortex shedding mechanism downstream of the lamps was visualised by the dye measurements. The residence time distributions obtained by measuring the breakthrough of a peak release of salt water confirmed the occurrence of recirculation zones, and showed that the mixing in the UV reactor was insufficiently at the lowest flow rates (resulting in laminar flow). The velocity profiles measured by laser Doppler anemometry also showed these recirculation zones. A main flow region in the middle of the reactor (in between the lamps) was distinguished, where the flow velocities were high, which caused the shortest residence times and lowest UV doses. The velocity profiles also showed the distortion of the flow pattern by the lamps, resulting in a wake region and large-scale vortices downstream of the lamp. The dose distributions obtained from the Lagrangian actinometry measurements showed a large peak in the low dose range, which was caused by the microspheres in the main flow region. A long tail in the high dose range was observed as well, caused by the microspheres trapped in the recirculation zone. Biodosimetry experiments were performed that determined the inactivation of *Bacillus subtilis* spores and MS2 phages. The wide dose distribution revealed that optimisation of the hydraulics (reducing the short-circuits and recirculations) is possible. The improvement of UV reactor performance is therefore explored in §6.4 by using CFD. The complex hydraulics in the UV reactor are a challenge for the CFD models. The validation of the CFD models by the different measurement techniques is therefore explored in the next section (§6.3).

6.3 CFD MODELLING OF A BENCH-SCALE UV REACTOR

6.3.1 Modelling conditions

The bench-scale reactor was modelled with both CFD models: the k-ε model and LES model. For the k-ε model, a mesh consisting of 161 719 elements was used to calculate the flow fields. Upstream of the reactor a 5 m pipe was modelled to obtain a fully developed turbulent velocity profile. A constant velocity was imposed at the inflow pipe. Downstream of the reactor a 1 m pipe was modelled with a zero pressure boundary. The closed boundaries were modelled as logarithmic walls.

For the LES model, a mesh consisting of around 8 million elements was used to calculate the flow fields. The inflow boundary condition was imposed from the resolved instantaneous velocity fields of another LES calculation, where a fully developed pipe flow was modelled. The fully developed turbulent flow was reached by imposing periodic inflow and outflow boundaries so that a pipe with infinite length was modelled. Downstream of the reactor a 0.5 m pipe was modelled with a zero pressure outflow boundary. The immersed boundary method was applied to resolve the complex boundary of the reactor (§2.2).

6.3.2 Mesh independency

The mesh independency was evaluated for both modelling approaches. The following conditions were imposed: a transmittance of 80%, a total lamp power UVC of 33.3 W, a flow rate of 5 m^3/h, and kinetic constants for *Bacillus subtilis* spores (Hijnen et al., 2006). For the k-ε model, the medium mesh in combination with 4900 particles was used as a reference. The prediced disinfection levels for the medium mesh size deviated around 1% from the fine mesh size (Figure 6.13), which was considered as sufficiently accurate. Also, increasing the number of particles resulted in a similar disinfection prediction. For the LES model, the medium mesh in combination with 50 000 particles was used as a reference. The coarse mesh almost showed the same disinfection levels (within 0.5% accuracy), Figure 6.14. Due to computational restrictions refining the mesh was not possible, so the medium mesh used the highest resolution possible. Considering the particles, a number of 50 000 particles was sufficiently accurate.

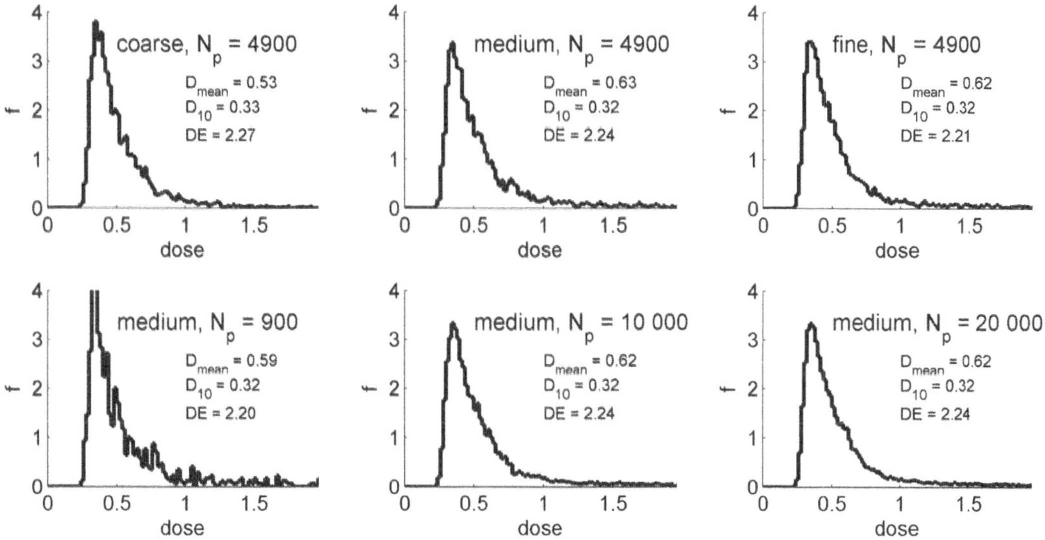

Figure 6.13 Dependence of the CFD calculations on the mesh size and number of particles for the k-ε model (coarse: 91 128 elements; medium: 161 719 elements; fine: 244 125 elements).

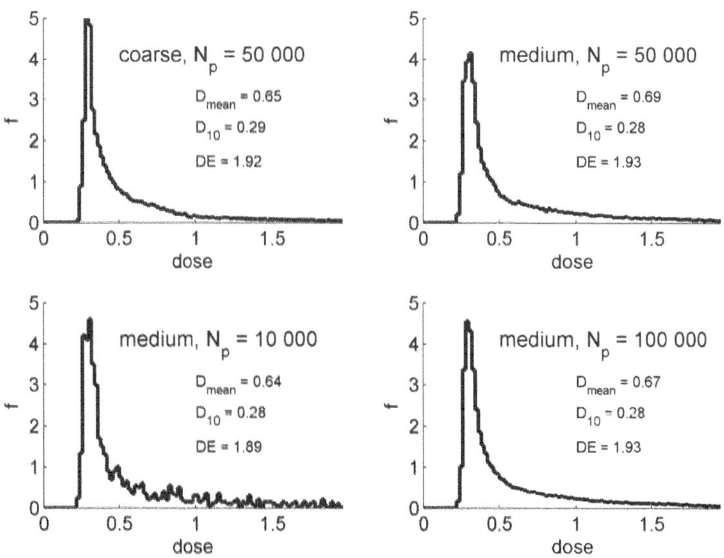

Figure 6.14 Dependence of the CFD calculations on the mesh size and number of particles for the LES model (coarse: 60 × 360 × 120 elements; medium: 90 × 480 × 180 elements).

6.3.3 Validation

The large data set of measurements for the bench-scale reactor allows a thorough validation of the CFD models. The validation was conducted for a flow rate of 5 m^3/h, except for the oxidation experiments where lower flow rates were necessary to increase the mean dose.

Residence time distribution measurements

The residence time distribution measured at the output sample port is compared to the computed residence time distributions (Figure 6.15). In the CFD models the total time to cover the distance between the measured input and output signals was recorded for each particle. The residence time distribution of the computed input signal was set equal to the measured input signal. The LES model predicted the residence time distribution well, whereas the *k*-ε model showed deviations with respect to the measurements. A narrower residence time distribution was predicted by the *k*-ε model: the shortest residence times were overestimated by the *k*-ε model, whereas the longest residence times were underestimated by the *k*-ε model. The deviation between *k*-ε model and measurement can be explained by an underestimation of the dead zones calculated by the *k*-ε model, whereas the lateral mixing was overestimated by the *k*-ε model. In the measurements and the LES model, the water trapped inside a recirculation region was less likely to be mixed with the main flow due to a lower degree of lateral mixing.

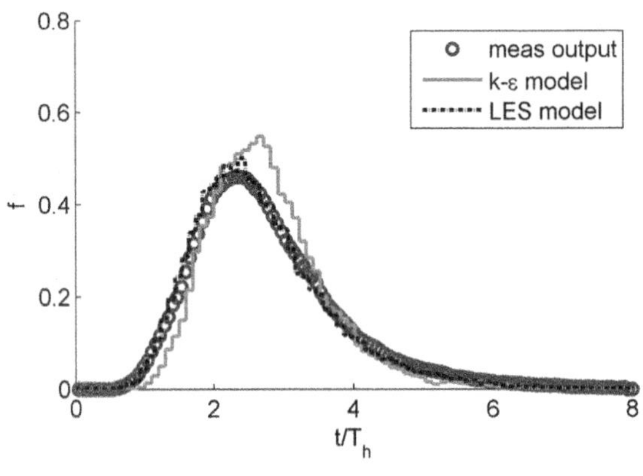

Figure 6.15 Measured residence time distribution compared to the *k*-ε model and LES model.

Flow fields

The computed and measured velocities (normalised with the average velocity in the pipe) are plotted in Figure 6.16. Four vertical lines in the middle of the reactor were considered. Upstream of the lamps the results of both modelling approaches agreed well with the measurements. Also, the main flow regions (with velocities equal to the mean velocity in the pipe) were well represented by both models. However, downstream of the lamps, the *k*-ε model started to deviate more markedly from the measurements (R^2 dropped significantly). The wake regions behind the lamps were underestimated by the *k*-ε model. Especially downstream of the upper right lamp, where the recirculation (negative velocities) stretched out over a long distance, the *k*-ε model did not predict negative velocities at all. The LES model showed a better agreement in the wake region: the negative velocities downstream of the lamps were well reproduced by the LES model.

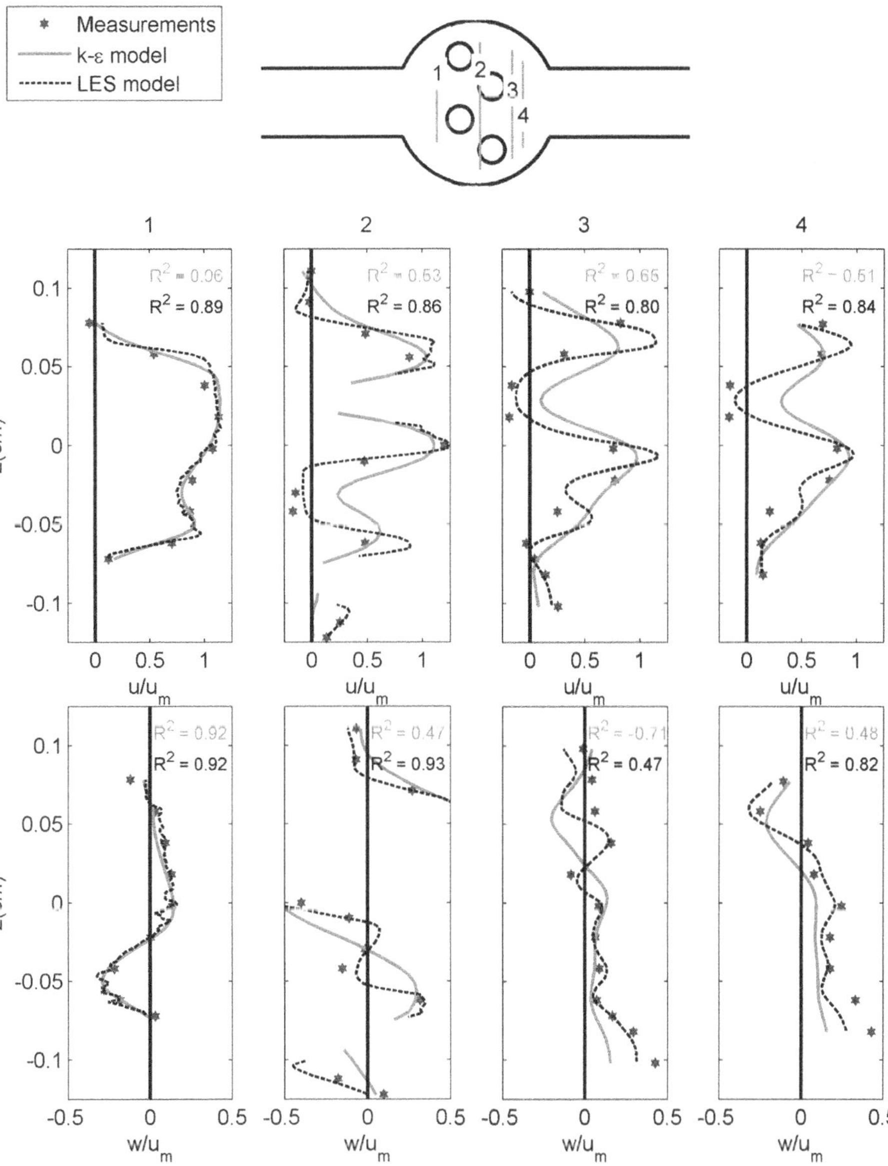

Figure 6.16 Measured flow fields (*u*, *w*) are compared to the computed flow fields of the *k*-ε model and LES model. The coefficient of determination (R^2), comparing the computed values with the measured values, is indicated for both models.

The vertical velocities in the wake region (Figure 6.16 lower panel) were also predicted incorrectly by the *k*-ε model. The upward velocities at the bottom of the reactor were underestimated, indicating that the large recirculation at the bottom was predicted too small. The LES model showed a better agreement here, showing that the recirculations in the top and bottom region of the reactor were well predicted.

UV irradiance

The UV irradiance was calculated by the irradiance model (§2.4.2) for the two different lamps (Table 6.1). First of all, the irradiance model was validated by experiments in air (Figure 6.17), which showed a good agreement between the measured and computed data. These measurements were conducted in a controlled area (Van Remmen UV Techniek), where the lamps could reach its optimal performance.

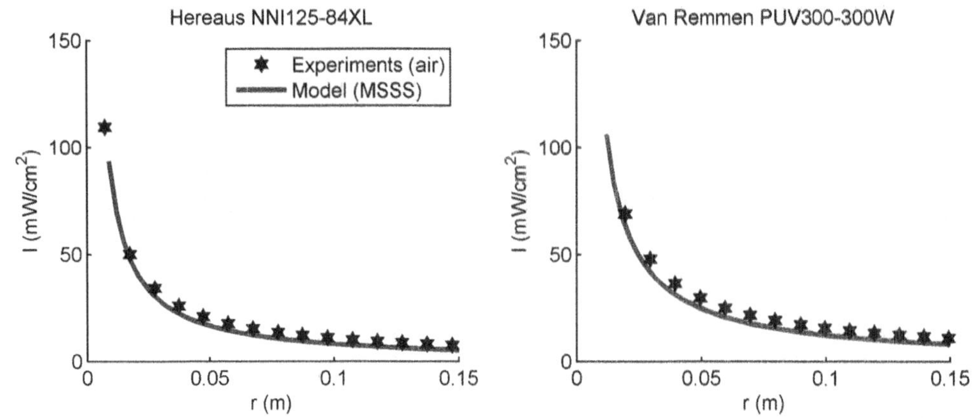

Figure 6.17 Measured versus computed UV irradiances in air for three lamp types (Van Remmen UV Techniek). The distance *r* is calculated from the centre of the lamp.

The lamp power and transmittance given in Table 6.2 were used to determine the UV radiation in the bench-scale reactor for the various experimental conditions. The point source segments of the lamp inside the reactor (over a length of 15 cm) were considered as well as segments outside the reactor. The latter represents UV that irradiated from the part of the lamp outside the reactor through the quartz tubes into the reactor. The quartz tubes, in which the lamps were installed, were relatively large in comparison with the lamp diameter, so that the UV radiation from the lamps outside the reactor could not be neglected. The temperature of the air surrounding the lamps could not be controlled in the experiment. Because of the temperature dependence of the lamp output and inhomogeneities of lamp output over the length of the lamp, an uncertainty of 10% was assumed for the lamp power. The uncertainty in the transmittance measurements by spectrophotometry was around 1%.

Lagrangian actinometry

The dose distributions calculated from the LES model and k-ε model were compared to the measured ones obtained from the Lagrangian actinometry experiment (Figure 6.18). The differences between the modelling approaches were most clearly visible at the low dose range, where the LES model predicted a higher peak, and at the high dose range, where the LES model predicted a longer tail. So, the LES model predicted a wider dose distribution than the k-ε model, which is in correspondence with the hydraulic measurements, showing a wider residence time distributions and more pronounced recirculations in the LES model. The mean doses were predicted around 5% higher by the LES model, mainly due to the larger recirculation zones and corresponding high UV doses in these zones. However, the median dose D_{50} was around 10–15% lower in the LES model, caused by the higher velocities in the main flow region. Nevertheless, the differences in dose distributions between both models were small. The overall shape of the dose distributions of both models agreed well with the measured data: they all showed a large fraction of particles in the lower dose range as well as the long tail in the high dose range. However, the measured first peak in the distribution (represented by the median dose, D_{50}) matched the LES model most closely in experiment 1, and matched the k-ε model most closely in experiment 3. For experiment 2 both models predicted a higher D_{50}, confirming the observation in §6.2.5 that the measured doses in experiment 2 seemed to be too low. The effect of the numerical error at the zero dose is investigated by manually removing this peak at the zero dose, which only leads to minor changes (the D_{50} increases to 56, 42 and 43 mJ/cm^2 for experiment 1, 2 and 3, respectively). Unfortunately, due to the variability in the measured dose distributions and the small differences between both models, the most accurate model could not be distinguished. However, the measurements showed that the most important flow features responsible for the large peak and long tail in the dose distribution were captured well by both models.

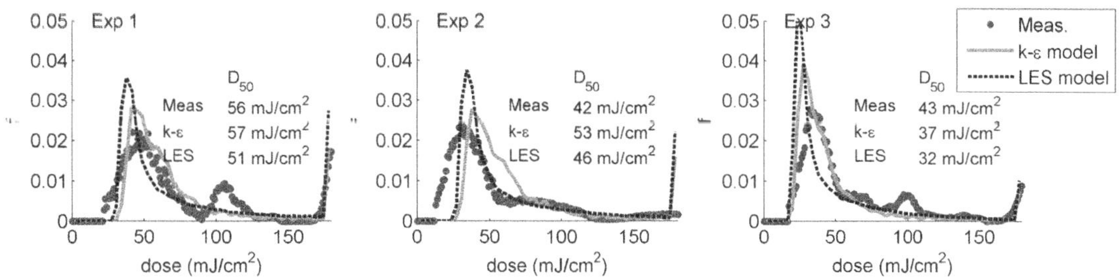

Figure 6.18 Dose distributions of the LES model and k-ε model compared to the measured distributions for the three experimental conditions. The distribution is characterised by the median dose (D_{50}).

Disinfection

The measured microbial inactivation of *Bacillus subtilis* spores and MS2 phages were compared to the CFD model results (Table 6.5). The LES model predicted on average a 10% lower disinfection than the k-ε model, which is explained by the larger short-circuit flows calculated by the LES model. These large short-ciruits flows caused a higher peak at the low dose range (observed in the dose distributions) and caused a lower disinfection. The inactivation levels predicted by both models were lower than the measured data, especially the LES model under predicted the measured inactivation levels. The models matched the measurements most closely in experiment 1, and deviated most pronouncedly in experiment 3. In Figure 6.19 the results of the biodosimetry tests are summarised. The collimated beam results and the dose-response curves from literature are displayed. For the CFD model results, a reduction equivalent dose (RED) was determined by calculating a single dose that corresponded to the calculated disinfection according to the dose-response curves from literature (ÖNORM, 2001; Hijnen et al., 2006). The RED is plotted against the measured inactivation in the bench-scale reactor, which must be in correspondence with dose-response behaviour in the collimated beam tests. An uncertainty area surrounding these data points is shown, accounting for the uncertainty in calculated irradiation (lamp power, measured transmittance) and in measured microbial inactivation (standard deviation of five samples). The computed microbial inactivations were a little more sensitive to UV than they should be according to the curves from literature and from collimated beam tests. Considering the variability in microbial inactivation of the collimated beam experiments and the uncertainty with respect to the lamp output (temperature dependence, inhomogenities in output over the length of the lamp), however, the over prediction by the CFD models seems to be small.

Table 6.5 The median dose measured by Lagrangian actinometry and microbial inactivation measured by biodosimetry are compared to the values calculated by the CFD models.

Exp #	$P_{reactor}$ (W)	$T_{10,w}$ (%)	D_{50} (mJ/cm^2)			*B. subtilis* spore*			MS2 phage*		
			Exp	k-ε	LES	Exp	k-ε	LES	Exp	k-ε	LES
1	31.9	87.7	**53**	57	51	**2.7**	2.6	2.2	**3.1**	2.7	2.4
2	46.9	78.4	**40**	53	46	**3.0**	2.3	2.0	**2.7**	2.5	2.2
3	31.9	78.5	**41**	37	32	**2.2**	1.5	1.3	**2.3**	1.8	1.6

*Log units.

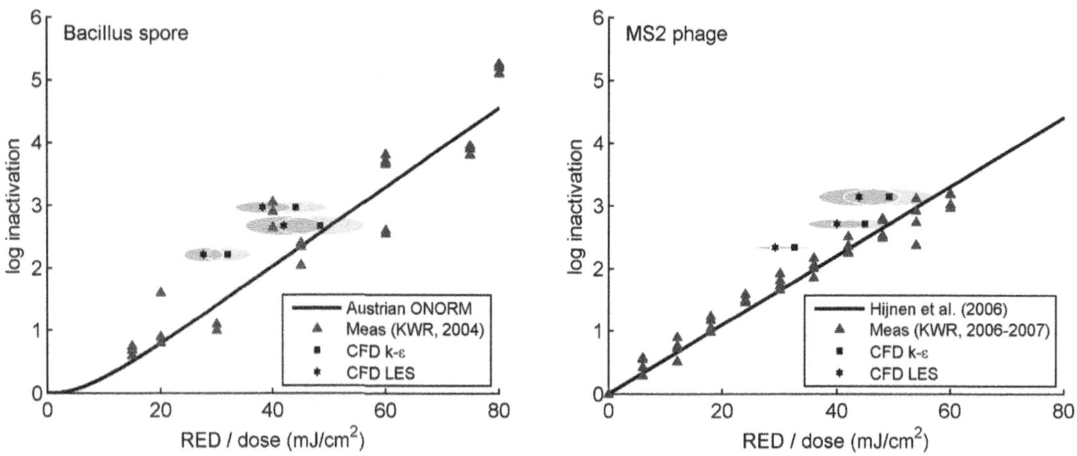

Figure 6.19 Dose-response curves from literature and from collimated beam tests. The RED calculated from the CFD models is plotted against the measured inactivation in the bench-scale reactor (the uncertainty bounds are visualised by the area surrounding the data points).

Oxidation

The measured degradation of atrazine is compared to the predicted values of both modelling approaches in Table 6.6. The background reaction rate of Dunea water fitted from the collimated beam data was used in the CFD models. Due to the low extent of atrazine degradation, the removal is less sensitive to the hydraulics, which explained the small differences between the modelling approaches. Both modelling approaches showed a good agreement with the measurements, indicating that the degradation of chemical substances can be predicted accurately by both models.

Discussion on the choice of the model

The differences between the LES model and k-ε model could be clearly seen from the averaged flow field: the spatial velocity gradients were larger in the LES model, whereas the k-ε model calculated a more smooth velocity field. As a result, short-circuits and recirculation zones were underestimated by the k-ε model, which is confirmed by LDA measurements and salt concentration measurements. Also, due to the differences in hydraulics, the LES model predicted larger fractions of particles with low and high UV doses, in other words, the dose distribution was wider. The LES model predicted a 10% lower disinfection than the k-ε model, which was less dramatic than the differences for the single UV lamp in Chapter 5. The bench-scale UV reactor is more confined, very large recirculations can not develop, especially those that are close or directly downstream of the lamp. The effect of underestimation of the recirculations by the k-ε model remains therefore limited. Trading off this accuracy with computation times, we have chosen to use the k-ε model to assess a large group of reactor designs in the next section. Moreover, using a LES model to assess all the reactors in §6.4 would require a huge effort due to the complexity of the different reactor geometries and large spatial domains that need to be resolved.

Table 6.6 Results of the oxidation experiments compared to the CFD models.

Flow rate (m^3/h)	Peroxide (mg/L)	D_{mean} (mJ/cm^2)		Atrazine removal (%)†		
		k-ε	LES	Meas	k-ε	LES
2.1	0	306	323	**21**	21	23
2.1	10	306	323	**43**	39	40
3.0	0	212	224	**14**	16	17
3.0	9.4	212	224	**29**	29	30

†Lamp power UVC: 88 W, transmittance water: 78%, initial concentration atrazine: 2 μg/L, background reaction rate: $\sum k_i N_i = 5 \cdot 10^4\ s^{-1}$.

6.4 DESIGN OF HYDRAULICALLY OPTIMISED UV REACTORS

An assessment of various geometries of UV systems is made to clarify how different types of reactors perform with respect to the inactivation of micro-organisms as well as to the oxidation of chemical compounds. These types include conventional systems with UV lamps placed perpendicular as well as parallel to the flow; annular systems with a varying lamp diameter; systems with a hydraulically optimised lamp shape (wing shape or grid in front); systems with a helical flow swirling around the lamp; and systems with alternating orientations of the lamps. A CFD model was constructed for each reactor type, from which the dose distribution was determined. The efficacy, in terms of disinfection or oxidation, could then be predicted. From these calculations, an estimate is obtained for the total energy consumption (for hydraulics and lamps) to achieve a certain extent of removal.

For disinfection, the removal of *Bacillus subtilis* spores is considered, for which the kinetic constants are given in Table 2.1. For oxidation, the degradation of atrazine with an injection of 10 mg/L peroxide is considered (constants given in Table 2.2). A background reaction rate fitted from collimated beam test with Dunea water was used (§6.2.7). All the reactors were subjected to a flow rate of 0.005 m^3/s and a UV transmittance in water of 80%. Total UVC power of all the lamps in the reactor was set to 200 W for disinfection and 2000 W for oxidation.

6.4.1 Overview of UV reactor designs

Four groups of reactors were considered (Figure 6.20): reference reactors, annular reactors, cross-flow reactors and streamlined lamp reactors. The reference reactors were used to show how the commonly used types of reactors perform. The other reactor types include new or existing reactor designs, sorted in a group of annular reactors, cross-flow reactors and streamlined lamp reactors. The last group shows the effect of hydraulic measures that optimise the shape of the UV lamp.

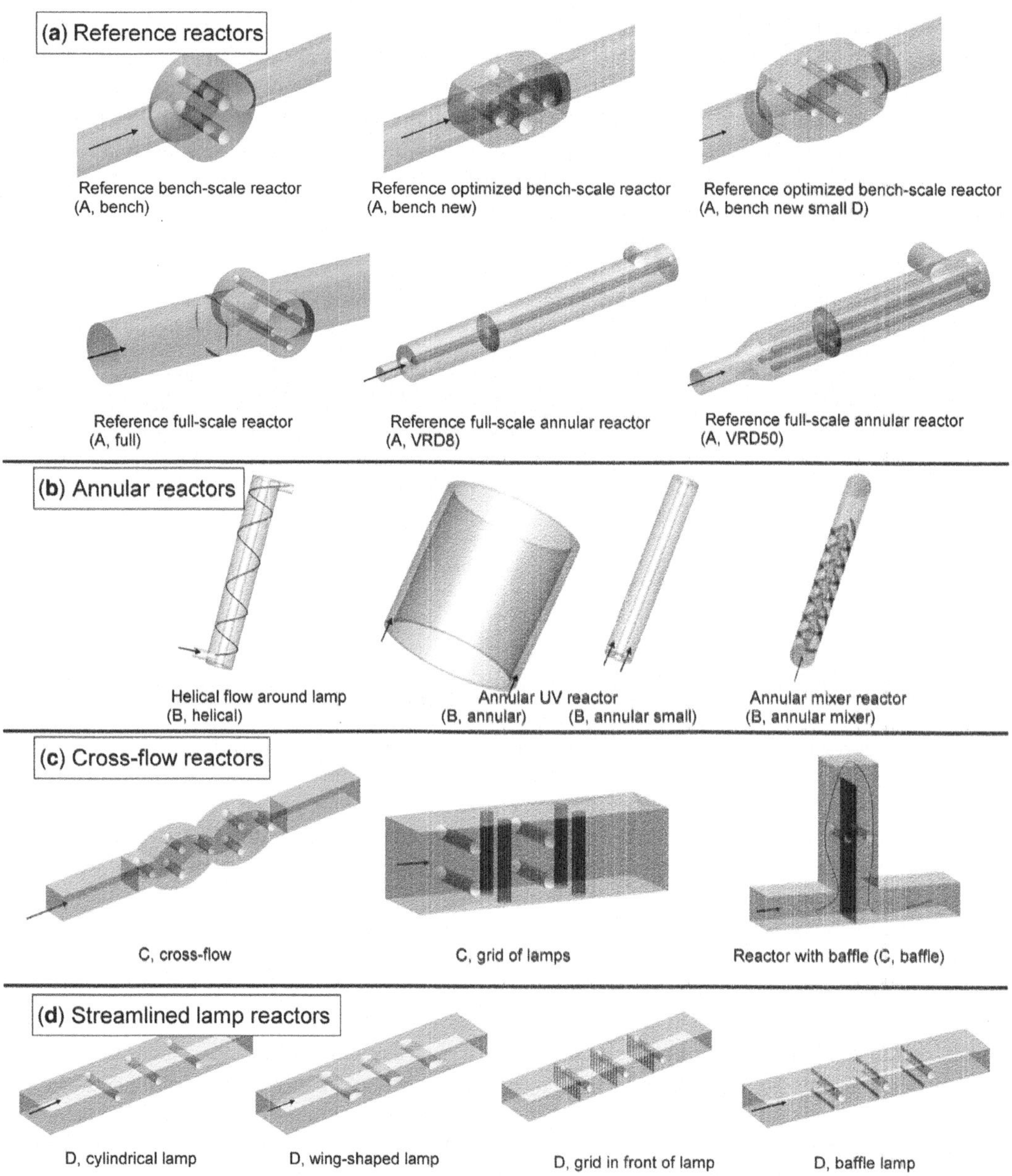

Figure 6.20 Overview of the geometries of the different UV reactors that were assessed by CFD.

Reference reactors

A few full-scale reactors were taken as a reference: a four-lamp perpendicular system (Rokjer et al., 2003), a one lamp annular system (VRD8) as well as a four-lamp annular system (VRD50), both developed by Van Remmen UV Techniek. Also, a few UV reactors used in research were simulated and served as reference cases. The bench-scale reactor previously used for validation purposes was simulated. Since suboptimal hydraulics were recognised in the bench-scale reactor from the measurements (§6.2), an optimised version of the bench-scale reactor was also simulated. Here, the upper and lower extremes of the reactor were cut off, so that the reactor was better streamlined with respect to the inflow and outflow pipes. Also, the lamp positions were changed. The optimised version was further improved by reducing the lamp diameter to 2.5 cm.

Annular reactors

Some other variants of annular systems were considered for calculation as well. The effect of the lamp diameter was investigated first. A system with a large lamp diameter was considered, one where the irradiance field is more uniformly distributed over the cross-section resulting in a narrower dose distribution. The smaller irradiances are compensated by the lower mean velocity in the system. The water flows parallel to the UV lamp. The outer diameter is equal to 0.95 m, whereas the diameter of the lamp equals 0.85 m. So, the distance from the lamp surface to the outer wall is equal to 0.05 m, which was found to be the optimal distance for a transmittance of 80% (§2.7.2). Also, a system with a smaller lamp diameter was considered, denoted as 'annular small'. The outer diameter is equal to 0.15 m, whereas the diameter of the lamp equals 0.05 m. Moreover, a helical flow system was considered, where the water is forced to circulate around the UV lamp. The swirling effect was achieved by the asymmetrical inflow and outflow pipes. Lastly, a static mixer with an annular lamp in the middle was considered, denoted as 'annular mixer'. Oblique baffle plates increase the amount of mixing in the system. For these three annular systems, the effect of a mirror layer on the outer wall was also investigated. Bolton (2000) indicated that stainless steel reflected less than 20% of the radiation in the UVC region, whereas Sommer et al. (1996) reported reflection rates of 60–65% for aluminium. So, it was assumed that 50% of the radiation was reflected by the mirror layer, whereas for the other calculation there was no reflection at all.

Cross-flow reactors

A number of cross-flow systems were also considered. In the first system, denoted as 'cross-flow', a number of perpendicular lamps were placed at a staggered pattern to force all the water parcels to move close to at least one lamp. The outer geometry of the reactor was chosen such that the distance from the lamp to the wall equals 0.05 m. Also, a system with a grid of lamps ($D_{lamp} = 0.05$ m) was investigated, denoted as 'grid of lamps'. Here, the axial direction of the lamp alternated to increase the amount of mixing in the system. The reactor consists of a square duct (0.3 m by 0.3 m), where the distance between the lamps equals 0.1 m and the distance from the lamp surface to the outer wall again equals 0.05 m. Lastly, a baffle was placed in the middle of the system to force the water to flow closer to the lamps. A lamp ($D_{lamp} = 0.05$ m) was placed in each compartment.

Streamlined lamp reactors

In these systems, the effects of measures that optimise the hydraulics around a perpendicular UV lamp (Chapter 5) were evaluated. Three UV lamps in a series were considered, each with a diameter of 0.05 m. The distance from the lamp surface to the outer wall was 0.05 m, similar to the annular system. First of all, cylindrical lamps were considered as a reference, denoted as 'cylindrical lamp'. Secondly, the cylindrical lamps were replaced by wing-shaped lamps, denoted as 'wing-shaped lamp'. Thirdly, a grid was placed upstream of each lamp to increase the mixing and reduce the wake behind the lamp, denoted as grid in front of lamp. Finally, baffles (height of 0.025 m) were added to force the main flow to move closer to the lamp and increase the mixing, denoted as 'baffle lamp'.

6.4.2 UV dose distributions

The prediction of the mean UV doses are presented in Figure 6.21 for all reactor types. The mean doses were normalised with respect to a characteristic dose ($D_{scale} = 132\,mJ/cm^2$ for a total power of 200 W, explained in §2.7.1). The mean dose was determined by taking the average of the doses of all the particles. The values were of the order of 0.55 to 0.9 of the characteristic dose. So, the amount of radiant energy absorbed by the flow differed up to a factor of 1.5 for the various systems. These differences were caused by: differences in lamp surface area, where radiant energy was lost by reflection and absorption; differences in distances to the outer wall, where radiant energy was lost to the outside of the system; and, differences in hydraulics. For the reference systems (group A), the modifications to the bench-scale reactor by cutting off the upper and lower parts, as well as reducing the lamp diameter, resulted in an increase in mean dose. The full-scale reactor showed a large mean dose, whereas the full-scale annular reactors (VRD8 and VRD50) had a smaller one. For the annular systems (group B), the helical system had the largest mean dose, indicating that the energy transfer from the lamp to the water was most efficient here. Due the high mixing and helical flow patterns, no short-circuit flows with high velocities and low doses were present here, resulting in a high mean dose. Increasing the lamp diameter for the parallel flow system resulted in lower UV irradiances in the water, so that the mean dose was smaller. Placing mirrors on the side walls (50% reflection) yielded an increase in the mean dose of around 15%. For the cross-flow systems (group C), mean doses were in general higher than for the annular systems without reflection. The streamlined lamp systems (group D) also showed high mean doses, except for the wing-shaped system, where less energy was transferred due to the larger lamp area.

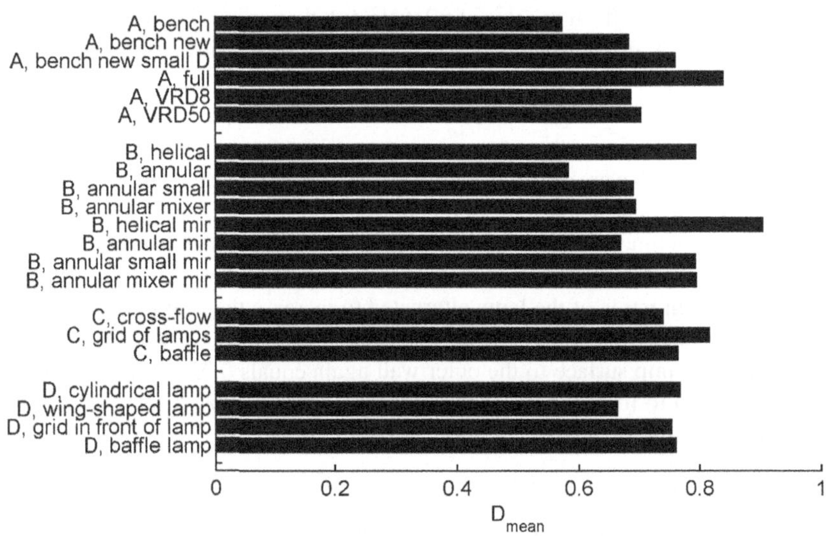

Figure 6.21 Prediction of the mean dose for all UV reactors by the CFD model. Mean dose is normalised with a characteristic dose (D_{scale}), §2.7.1.

The results of all the UV systems are summarised in Figure 6.22, where the dose distribution for each system is plotted. Also, the normalised mean and minimal doses, the log inactivation of *Bacillus subtilis* spores (*DE*) as well as oxidation of atrazine (*OX*) are given. The dose of the particle with the lowest dose defined the minimal dose. In Figure 6.23, the disinfection and oxidation results for each reactor are plotted as well. The various UV reactors showed the following results:

- For the reference reactors (group A, Figure 6.22a-f), the bench-scale reactor showed a peak in the dose distribution at a small dose, which were caused by the short-circuit flows that moved in between the lamps, resulting in small inactivation levels. The geometric modifications to the bench-scale reactor reduced the short-circuits flows (particles with high velocities that had moved in between the lamps) and resulted in a shift of the peak in the dose distribution towards a higher dose. The reference full-scale reactor showed high oxidation levels as well as high disinfection levels, mainly because of the high mean dose. For the annular reference systems, the four-lamp system (VRD50) showed a narrower dose distribution than the one-lamp system (VRD8), because the UV irradiance field is more uniform for the four lamp system. The disinfection levels for this system were the highest of all the reference systems.
- For the annular reactors (group B, Figure 6.22g-j), the annular mixer and helical system showed a nearly symmetrical dose distribution. These systems showed the best oxidation results. The high intensity of mixing in the annular mixer system had resulted in a very narrow dose distribution. Consequently, this system showed the highest disinfection results. The other annular system had asymmetrical dose distributions with a peak at the lower dose part. The peak was related to the particles travelling close to the outer wall of the reactor. The annular system with the large lamp diameter showed a smaller dose distribution than the system with a small lamp diameter, as expected, however the mean dose was smaller, because more UV radiation was lost due to the larger quartz sleeve. The overall effect of increasing the lamp diameter was a slightly better disinfection, but a smaller oxidation performance.
- Placing mirrors at the outer walls (Figure 6.22k-n), such that 50% of the UV radiation is reflected, shifted the peak in the dose distribution to higher UV dose levels. Since the reflected radiation had the greatest contribution near the outer wall, where the particles with the lowest doses were located, the smallest doses greatly increased. As a consequence, the disinfection and oxidation performance increased considerably. The annular mixer system with a mirror showed the best disinfection performance of all the systems, whereas the annular helical system and the mixer system showed the best oxidation performance.
- For the cross-flow reactors (group C, Figure 6.22o-q), the oxidation levels were predicted high because the mean doses were high. However, the dose distributions were very wide, because a part of the bulk flow moved through areas with much lower irradiation levels than the levels close to the lamp. As a result, especially for the grid of lamps, the disinfection levels were predicted low.
- For the streamlined lamp reactors (group D, Figure 6.22r-u), the reactor containing the cylindrical lamps had a large peak at the smaller doses, so that the disinfection performance was very low. The peak in the dose distribution at the lower doses was caused by the particles travelling close to the outer walls. The low irradiance intensities in combination with high velocities near the outer walls resulted in low doses. The oxidation results were reasonably good because the mean doses were high. The peak in the dose distribution could not be reduced by changing the lamp to a wing-shaped lamp. Placing a grid in front gave a small reduction in the peak of the dose distribution, so that the disinfection was slightly increased. Adding baffles close to the lamps had a major impact: the dose distribution became more symmetrical and the peak shifted towards a higher dose. The baffles forced the particles to move closer to the lamp, so that the particles close to the outer wall responsible for the low doses would receive higher doses. As a result, log inactivation was almost doubled.

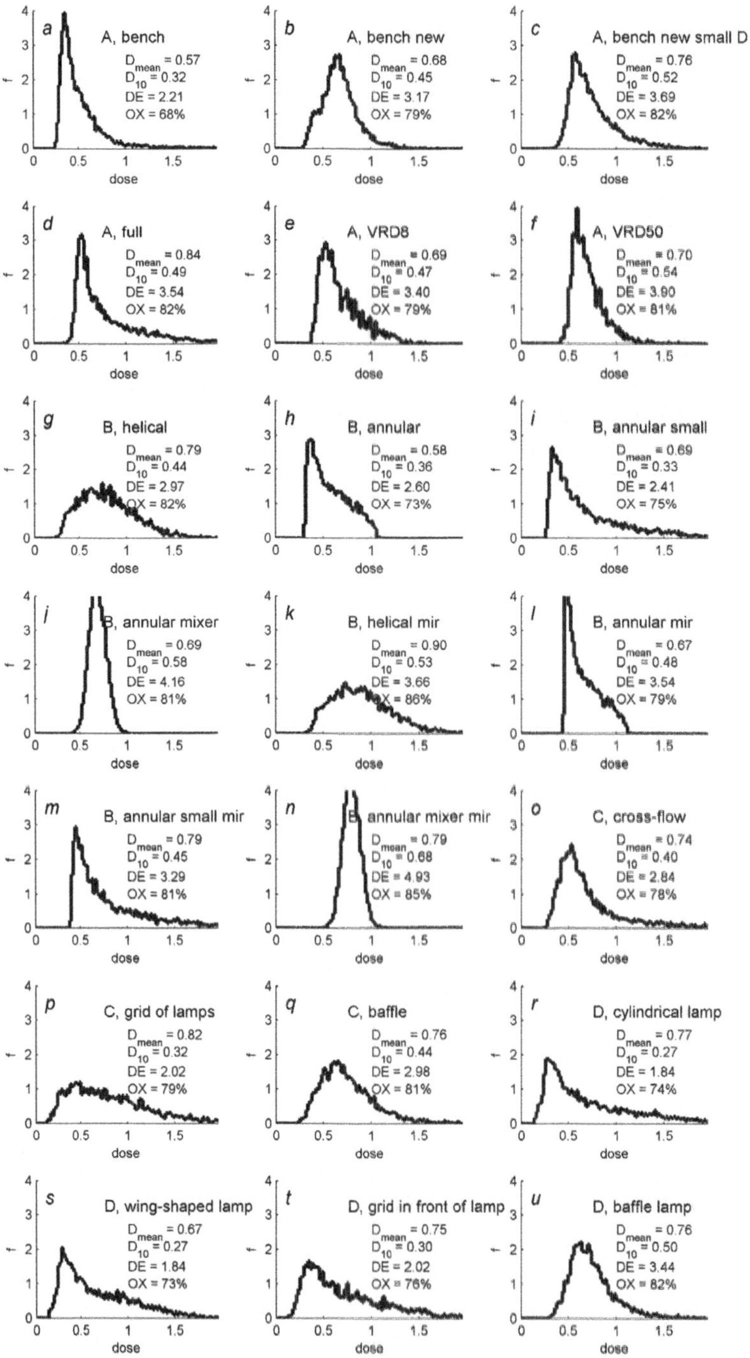

Figure 6.22 Dose distributions (normalised) for the different reactor designs. Parameters shown are the normalised mean dose (D_{mean}), normalised 10th percentile dose (D_{10}), log inactivation of *Bacillus subtilis* spores (*DE*) and degradation of atrazine (*OX*).

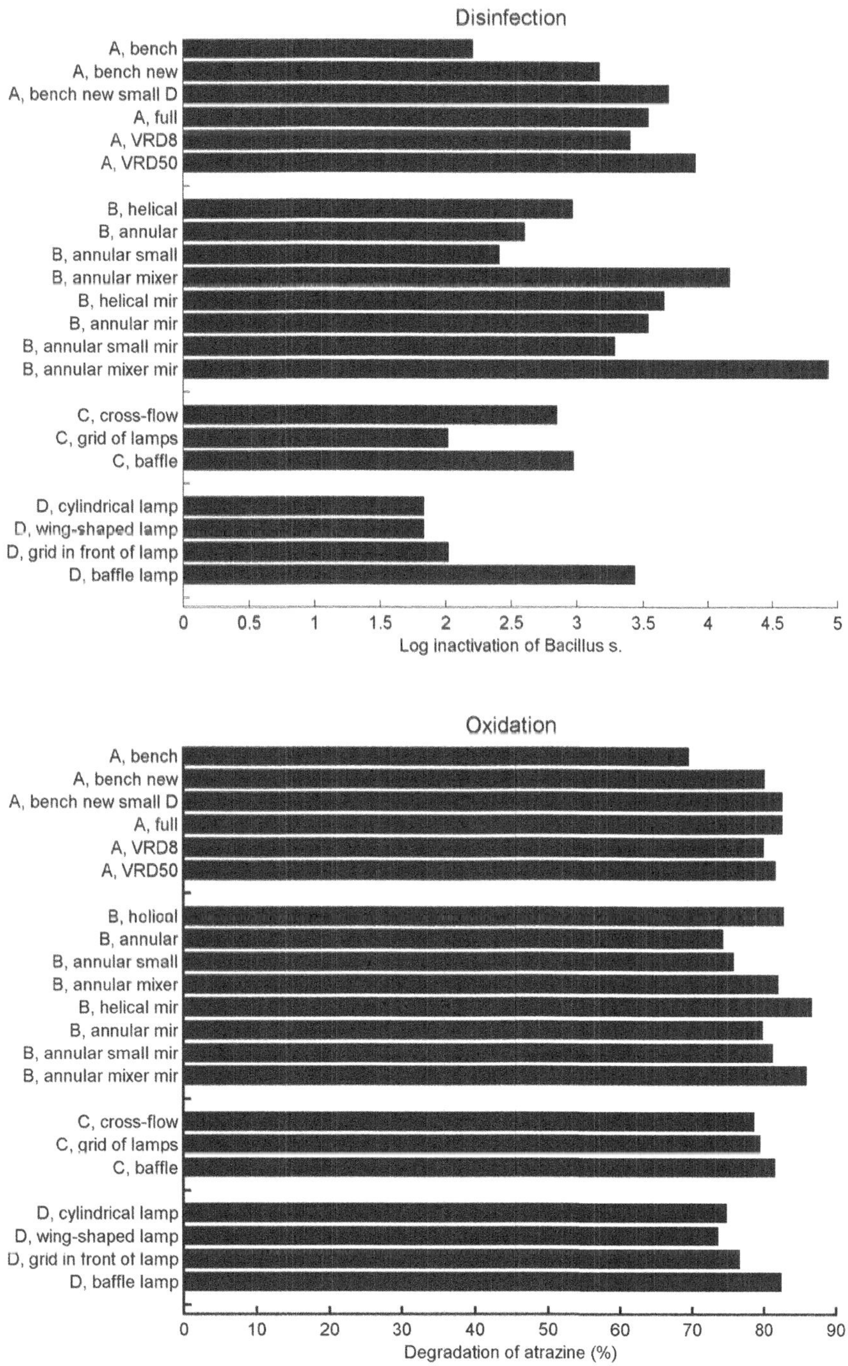

Figure 6.23 Prediction of the disinfection of *Bacillus subtilis* spores (upper panel) and oxidation of atrazine in combination with peroxide (lower panel) for each UV reactor.

An overview of the disinfection and oxidation results of the best performing reactor in each group is shown in Table 6.7. The reference systems performed well, because they were already a result of hydraulic optimisations, which are clearly visible for the bench-scale reactor, where the modifications had resulted in an increase in inactivation levels of almost 100%. But still the annular mixer showed a big improvement in disinfection performance with respect to the reference systems. By placing a mirror at the outer wall, the disinfection performance was even higher. For the removal of chemical compounds, differences between the systems were less pronounced because oxidation is less sensitive to the dose distribution. But still the annular mixer showed an improvement for the degradation of chemical compounds.

Table 6.7 Overview of the results of the disinfection and oxidation for the best design of each UV reactor group.

Group	Best design	Disinfection†	Oxidation
Reference	A, VRD50	3.90	81%
Annular	B, annular mixer	4.16	81%
Annular mirror	B, annular mixer mirror	4.93	85%
Cross-flow	C, baffle	2.98	81%
Streamlined lamp	D, baffle lamp	3.44	82%

†log units.

The results of the disinfection and oxidation predictions are plotted against the mean dose and the 10th percentile of the dose (D_{10}) in Figure 6.24. The D_{10} characterises the low dose part of the distribution better than the minimal dose because it accounts for the skewness of the dose distributions. A strong correlation between the D_{10} and the disinfection ($R^2 = 0.99$) was observed, showing that the D_{10} largely determined the disinfection. The slope of the fit matched with the inactivation constant of *Bacillus subtilis* spores used here. The correlation between the mean dose and the disinfection was weak ($R^2 = 0.07$). For oxidation both the D_{10} as the mean dose showed a moderate correlation with the degradation. The differences between oxidation and disinfection are mainly related to the extent of removal. For disinfection, the removal of micro-organisms reaches values between 99% (log 2) and 99.999% (log 5), which means that the lower doses are very critical for disinfection. For oxidation, the removal of organic compounds reaches values between 60% and 85%, which are less sensitive to lower doses.

6.4.3 Results of energy consumption

The total use of electrical energy is shown in Figure 6.25. The energy consumption consists of electrical energy use by the UV lamp (assuming an efficiency of the lamp of 30%) and hydraulic loss (assuming a pump efficiency of 50%). The differences in hydraulic energy losses between the systems were large, which was mainly caused by the differences in velocities. For the VRD8 reactor, for example, a higher flow rate was applied than the design conditions of the reactor would allow. To that end, maximum velocities of around 4 m/s occurred in the holes of the baffle plate, so that the pressure loss was very high. Under the design conditions, pressure losses will be comparable to the VRD50 system. The inflow conditions in the helical flow reactor (A, helical) were responsible for a large pressure loss due to a sudden lateral expansion. In general, for disinfection, the hydraulic losses were small with respect to the lamp power. For oxidation, the hydraulic losses were negligibly small with respect to the high lamp power. Because the hydraulic energy losses are small with respect to the lamp energy, hydraulic measures that increase mixing are of great interest. Therefore, the total power consumption is lowest for the annular mixer system.

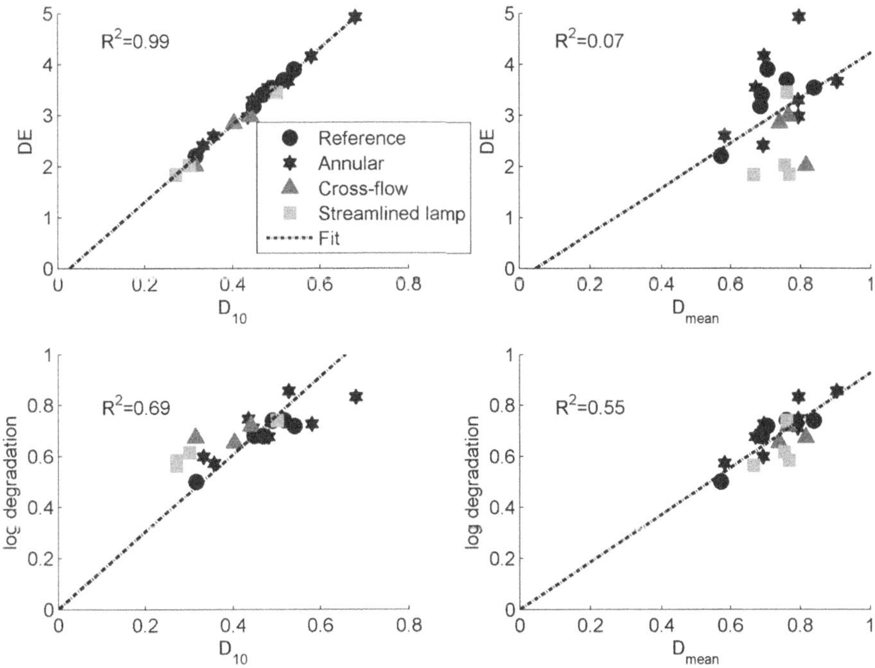

Figure 6.24 Disinfection (upper panels) and oxidation (lower panels) results as a function of D_{10} (dose at which 10% of the particles has a lower value) and D_{mean} for each UV reactor. D_{10} and D_{mean} were normalised with the characteristic dose.

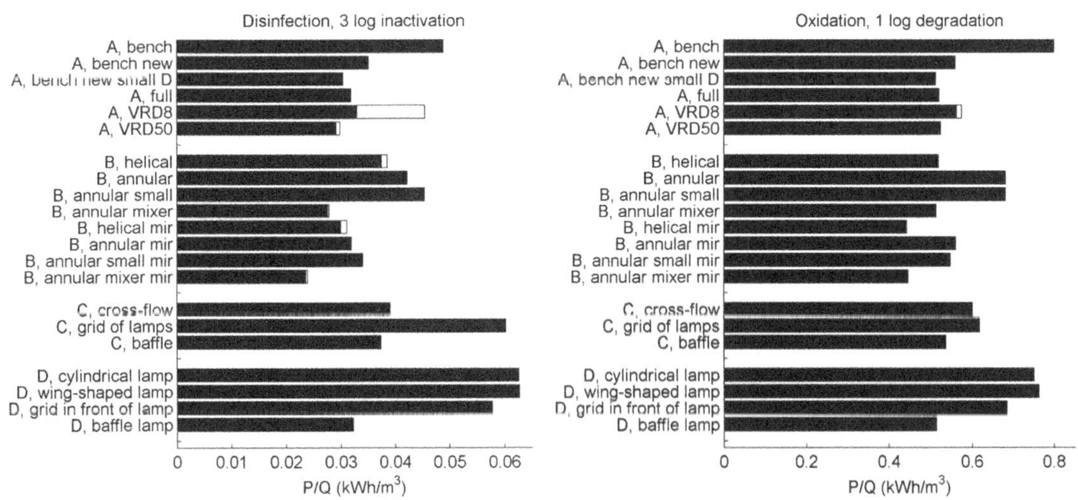

Figure 6.25 Electrical energy consumption by UV lamps (black bar) and hydraulic losses (white bar) needed to obtain 3 log disinfection of *Bacillus subtilis* spores (left) or 1 log degradation of atrazine (right). The white bars are in some cases very small (not visible), indicating that the hydraulic loss is negligibly small.

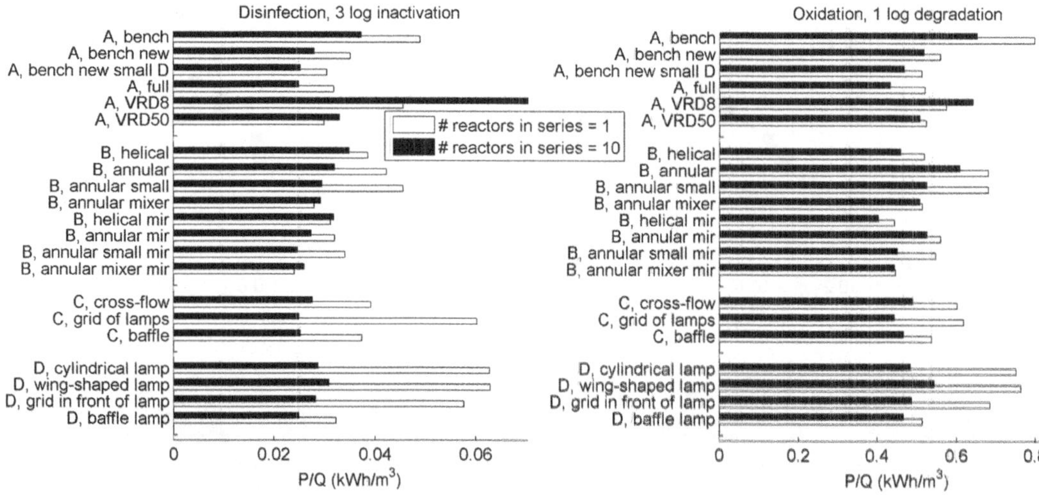

Figure 6.26 Total energy consumption needed to obtain 3 log disinfection of *Bacillus subtilis* spores (left) or 1 log degradation of atrazine (right) for 1 reactor and 10 reactors in series.

In Figure 6.26, the effect of placing a number of reactors in series is shown. The radiant power for the system with n reactors in series is reduced by a factor of n in comparison with the system with 1 reactor. In between each reactor in series, a static mixer is assumed, so that the particles are redistributed over the cross-section. According to the central limit theorem, the dose distribution becomes smaller and approaches closer to the normal distribution after each reactor. As a consequence the disinfection and oxidation results of the system must increase. However, the hydraulic losses will also increase by a factor of n. The total energy consumption needed to obtain 3 log inactivation of *Bacillus subtilis* spores or 1 log degradation of atrazine is therefore plotted in Figure 6.26 in the case of $n = 1$ and $n = 10$. For disinfection, the differences in energy consumption between the systems became smaller when a number of reactors were placed in series. The systems with suboptimal hydraulics (such as 'A, bench' or 'D, cylindrical lamp') had improved considerably. However, the systems with optimal hydraulics (such as 'B, annular mixer' or 'B, helical mir') did not show an improvement because the disinfection was only slightly increased by placing reactors in series, whereas the hydraulic power losses were increased as well. For oxidation, the relative gain in energy consumption by placing 10 reactors in series was less than for disinfection, but the absolute gain in energy was much higher. For almost all the systems the energy consumption can be improved by placing reactors in series because the hydraulic power losses are negligibly small (except for the 'A, VRD8'). However, placing reactors in series is most favourable for reactors that show suboptimal hydraulics.

In Figure 6.27, the factor E/E_{ideal}, interpreted as the additional energy needed to obtain the same log inactivation or degradation as for the ideal situation, is plotted against D_{10}/D_{mean}. The energy consumption E was determined from the power consumption of the UV lamps required to obtain a 3 log inactivation or 1 log degradation using the resolved dose distribution from the CFD model. The ideal energy consumption E_{ideal} is determined similarly, only for a dose distribution with a peak at the mean dose. The kinetic constants given in Table 2.1 and Table 2.2 were used. The values from the CFD model coincide well with the analytical model derived from the exponential dose distribution for disinfection as well as oxidation. So, the exponential dose distribution as described in eq. 2.32 and 2.33 can be used to predict the energy consumption of a UV system. In this way the analytical model forms an easy tool to investigate the energy reduction that could be obtained when the D_{10}/D_{mean} is increased. Moreover, it allows for prediction of the sensitivity of the energy reduction with respect to the kinetic constants. The maximum energy reduction is obtained when the dose distribution is reduced to $D_{10} = D_{mean}$. The figure clearly indicates the potential in reduction of energy for the various systems. The annular mixer system showed a E/E_{ideal} close to 1, so that this system is hydraulically almost perfect: little reduction in energy can be obtained by further reducing the dose distribution (10%). The D_{10} for existing UV reactors was of the order of 0.5 to 0.75 of the mean dose, so that a hydraulic optimisation would result in an energy reduction from 15% to 50% for disinfection. For one of the reference systems (VRD50), the hydraulics were close to the ideal situation (D_{10}/D_{mean} of around 0.75), so that the potential in reduction of energy is small (maximum of 15%). For other systems with a D_{10}/D_{mean} of around 0.5, the energy reduction induced by hydraulic optimisations could be larger.

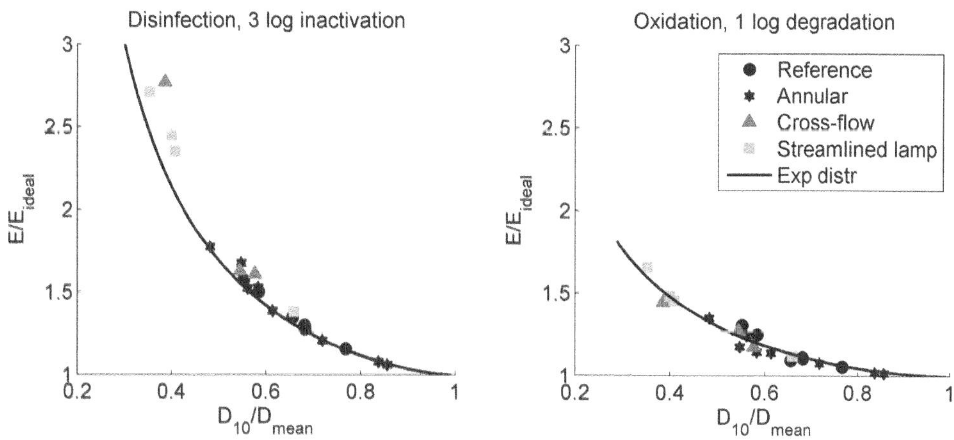

Figure 6.27 Energy consumption of UV lamps is plotted against the dose distribution. The energy consumption is shown as the calculated energy E divided by the E_{ideal} in the case of perfect hydraulics. The dose distribution is characterised by the 10th percentile dose divided by the mean dose.

6.4.4 Design considerations

The combination of particle positions and particle velocities determine the dose distribution. The particle positions determine the amount of UV radiation that is received by the particle, which can differ largely due to the large spatial differences in UV radiation field. So, the weak points in the UV systems were related to the parcels of water that travel at the greatest distance from the lamps. At these locations, usually close to the walls, irradiance intensities were low due to absorption of UV radiation in water. For example, the weak points in the annular system were related to the particles that remained close to the outer wall. The system with a helical flow around the lamp or annular mixer therefore had a higher disinfection efficacy due to increased mixing. As a result, particles with low doses did not remain at the outer wall, but also travelled closer to the lamp surface. For the cross-flow systems, particles travelling close to the outer wall were also responsible for the low disinfection levels. Similarly, particles that remained in the corners of the system with alternating lamp orientations caused a wide dose distribution. The optimised bench-scale reactor showed good results because the staggered lamp positions forced all the parts of the flow to travel close to at least one lamp. For a good reactor design, it is therefore important that all the particles reach the area close to the lamps. This can be achieved by increasing the mixing in the system and/or by manipulating the main flow, such as placing baffles. The mirrors used in the annular systems resulted in a significant increase in disinfection and oxidation, because the irradiance at the outer wall was increased by 50% (in the case of 50% of mirroring). The particles with low doses that had moved along the outer wall benefited from an increase in UV dose.

The effect of increasing the radius of the lamp is ambiguous: on the one hand, the irradiance field was more uniform, so that the dose distribution was narrower; on the other hand, the UV intensity was lower, so that the mean dose was reduced. For the annular system, the overall effect of increasing the lamp diameter was positive for the disinfection capacity. However, for the bench-scale reactor, the effect was the opposite: the system with the larger lamp diameter had a lower predicted disinfection.

6.5 CONCLUSIONS

A thorough experimental validation of a bench-scale UV reactor was performed, measuring flow fields, residence time distributions, dose distributions, microbial inactivations and chemical degradations. The hydraulic measurements show that recirculations occur in the upper and lower part of the UV reactor, resulting in long residence times and an inefficient distribution of UV doses. The extensive data set is used for validation of both CFD models (k-ε model and LES model). The differences between the modelling approaches were less pronounced than for the baffle configuration discussed in Chapter 5, mainly because the reactor was more confined, so that very large recirculations could not develop. The LES model showed a good resemblance to the measured velocity fields and residence time distributions, whereas the k-ε model showed some deviations, mainly at the recirculation zones. The biodosimetry and Lagrangian actinometry experiments were too inaccurate to allow a determination between the models. In the end, the LES model predicted a 10% lower disinfection than the k-ε model.

For the design of UV reactors the k-ε model was used to predict the performance of a wide range of UV reactors. The best results were obtained by an annular system, designed as a static mixer with a lamp in the middle, which came closest to the perfect hydraulic system (within 10%). The efficacy of UV systems can be improved by taking measures that increase the mean dose and/or narrow the dose distribution:

- Enhance the mixing inside the system to ensure that all the micro-organisms come close to the lamps and the dose distribution becomes narrower. This can be achieved by placing baffles or by forcing a helical flow around the lamp. Since the hydraulic energy loss is small with respect to the lamp energy, the additional hydraulic loss by increasing the mixing is beneficial.
- Place mirrors at the outer walls, especially in the case of an annular system, which leads to a significant increase in the mean dose and therefore in disinfection and oxidation levels.
- Place a number of reactors in series with a static mixer in between the reactors to narrow the dose distribution, which is most effective for reactors with suboptimal hydraulics.

Chapter 7

General conclusions

In the past, the water industry often neglected the (complex) hydraulics. However, in the last decade when the tools to describe the hydraulics became more widely available, we have seen more interest for research in this field. In this thesis, the importance of understanding the hydraulics for different water treatment processes was shown. Ozone and UV systems were evaluated by computational fluid dynamics (CFD), an approach that uses advanced numerical models to model flow, reaction and mixing in these systems. The models were extensively validated by different measurements techniques. Due to the complexity of the turbulent flows, different numerical approaches were evaluated that resolve the turbulent motions at different levels of detail. Using the CFD models gave better insights in the design aspects of water treatment installations. Conclusions related to different CFD modelling aspects as well as to the design aspects of reactors are presented here.

7.1 CFD MODELLING ASPECTS

Figure 7.1 displays all the steps required to predict the disinfection or oxidation performance of a UV reactor. A comparable approach can be followed for other applications, e.g., ozone systems. This scheme is constructed for the large-eddy simulation. The large-eddy simulation (LES model) resolves the large-scale time-varying turbulent motions, whereas the Reynolds-averaged Navier-Stokes (RANS) approach solves an averaged velocity field and models the turbulent fluctuations by means of an artificial viscosity term. Similar steps as in this scheme are followed for RANS models, except for the calculation of instantaneous fields of velocity (or concentration). First of all, the geometry and boundaries of the domain need to be defined carefully. The mesh is obtained by dividing the geometry into small elements. For the LES model, the mesh needs to be sufficiently fine to resolve the relevant length scales of the turbulent motions. After the boundary conditions are imposed, instantaneous velocity fields are computed. An averaged velocity field is calculated by averaging the ensemble of instantaneous velocity fields. Subsequently, the trajectories of passive particles are calculated by integrating the instantaneous velocity fields in the case of the LES model. In the RANS model, a random walk model is used to represent the turbulent fluctuations (Chapter 3). Also, the UV intensity or concentration of disinfectant (or oxidant) is calculated. The UV intensity is assumed to be of steady-state and is independent of the flow fields. The dose is calculated for each particle by integrating the disinfectant or oxidant (e.g., UV irradiance, ozone) over the particle's trajectory. The disinfection or oxidation can be calculated from the distribution of doses using the appropriate kinetic models. The concentration of micropollutants (micro-organisms or organic substances) can also be directly solved on the computational mesh (lowest panel in Figure 7.1). The LES model calculates an ensemble of instantaneous concentration fields, which are averaged to obtain the degradation performance, whereas the RANS model immediately calculates an averaged concentration field.

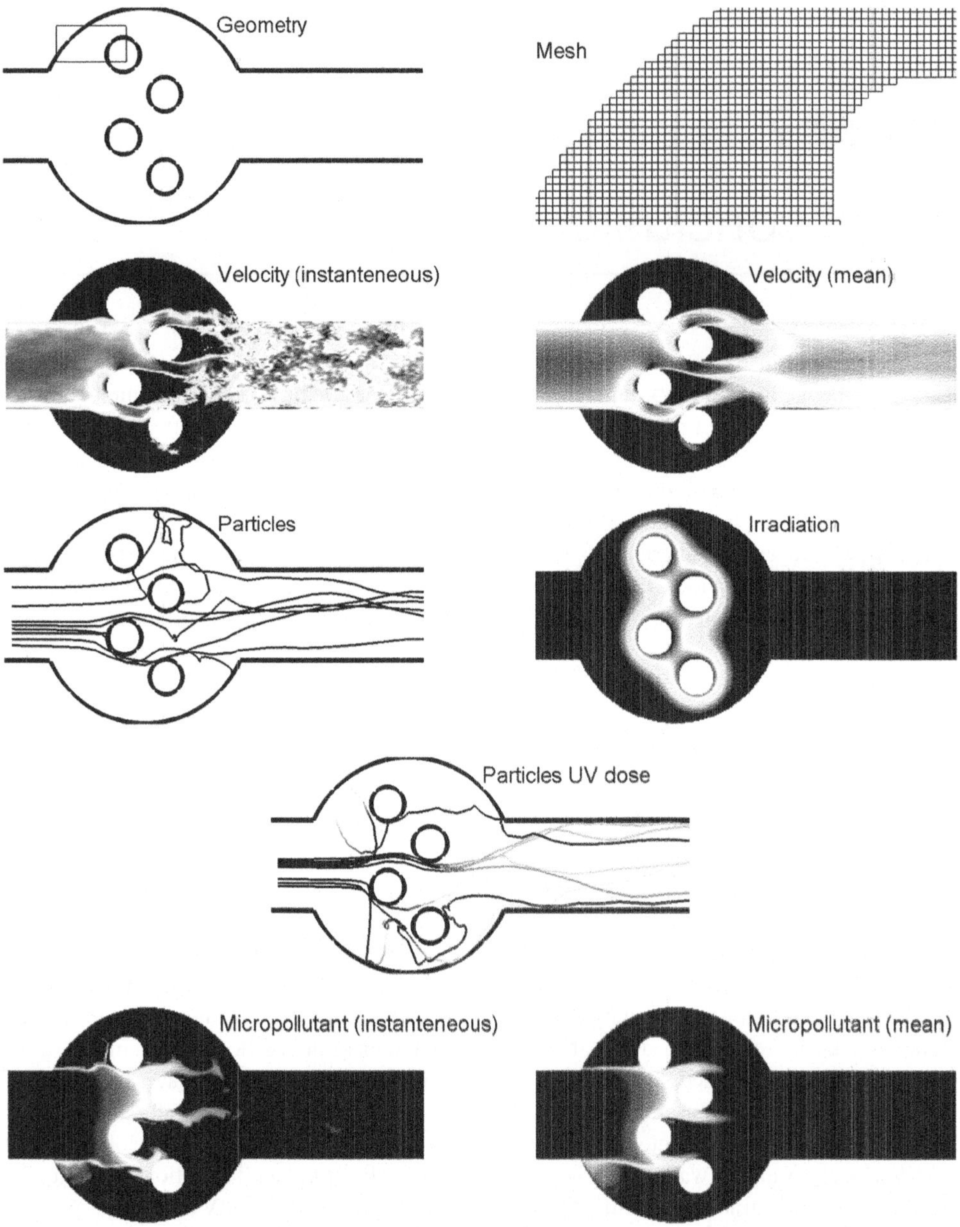

Figure 7.1 Overview of the necessary sub-models to predict the performance of water treatment systems. This figure shows modelling of a bench-scale UV reactor by the LES model. A movie can be found at: http://dx.doi.org/10.4121/uuid:474d9ab9-c354-4d18-87c9-69b02acedce3.

7.1.1 Turbulence modelling

The most challenging task of the CFD model is to model the turbulence. Resolving all the turbulent length and time scales is impossible for the present day computers, so simplifications are necessary. The RANS approach with a k-ε turbulence model is most often used in engineering practice. The averaged velocity field calculated by the RANS model will always differ significantly from the instantaneous velocity fields occurring in nature. The large-eddy simulation resolves more details of the turbulent flow. Both modelling approaches were assessed for a simplified cross-flow UV system and a bench-scale UV reactor. This assessment demonstrated the weaknesses of the RANS approach with a k-ε model. By simplifying the turbulent flow inside the UV system, flow fields are predicted inaccurately in time and space. Comparison of computed velocity profiles by experimental data confirmed that the LES model was more accurate than the k-ε model. In general the k-ε model predicted a more smooth velocity profile, showing smaller velocity gradients, so that short-circuit flows and recirculation zones were less pronounced than for the LES model. How much this affects the predicted disinfection depends on the local geometry and the spatial distribution of UV radiation. By adding baffles to a single cross-flow UV lamp, the predictions of the k-ε model became inaccurate in situations where large velocity differences are present and large wakes can develop. Especially in the cases where large short-circuits developed at a close distance from the lamp, the prediction of these short-circuits became relevant, and large differences in predicted disinfection were found between the modelling approaches. The latter may not be the case for full-scale installations, where the wakes are more confined due the interaction with other lamps. For the bench-scale UV reactor, the differences between both modelling approaches were indeed less pronounced. The LES model predicted less optimal hydraulics than the k-ε model: stronger short-circuits and larger recirculations were found, however, the differences in disinfection predictions were small (around 10%).

7.1.2 Lagrangian versus Eulerian

In principle both techniques, Lagrangian or Eulerian, could be used to predict the disinfection or oxidation performance in treatment systems. In the Lagrangian technique the movements of individual particles are solved, whereas in the Eulerian technique an advection-diffusion-reaction equation is solved on the computational mesh. Since both techniques use the same information from the resolved flow model, they should theoretically give the same results. For a cross-flow UV system it was shown that the oxidation predicted by a LES model was similar in the Lagrangian and Eulerian calculations. For the ozone systems it was shown that both techniques gave the same results for the residence time distributions as well as the disinfection, provided that sufficient particles are used in the Lagrangian approach. A number of 5000 particles was sufficient for a steady-state calculation used by the RANS model, whereas a number of 50 000 particles was sufficient for an unsteady calculation used by the large-eddy simulation.

The advantage of the Lagrangian technique over the Eulerian technique is that it provides valuable statistical information, since different variables can be collected for each particle. These statistics provide additional information about the performance of the installations, for example dose distributions. Another consideration for choosing a Lagrangian or Eulerian approach is computing time. The computations required for the Lagrangian approach are more time-consuming than for the Eulerian approach. Once the Lagrangian computation has been made, however, parameters in the kinetic model or radiation model can be adjusted much quicker, since it only involves a post-processing step, appointing values to the different properties of the particles. In the Eulerian approach the complete transport and reaction calculation of micro-organisms and/or chemicals has to be repeated.

7.1.3 Validation techniques

The assumptions made to model the turbulence require a thorough experimental validation of the CFD results, especially when complex flow patterns occur. In this thesis, various experimental techniques were explored to obtain a better understanding of hydraulic processes and to validate the models. The following techniques were used:

- Passive tracer tests determine the residence time distributions, which indicate the presence of recirculation zones and short-circuits. It can be used for validation, but the residence time distribution is an integrated property of the system, so the locations of recirculation zones and short-circuits can not be pointed out. Moreover, the residence time distributions characterise the hydraulics, but the total performance of the system also depends on the spatial distribution of disinfectant (or oxidant).
- Visualisation studies with dye show the (turbulent) flow patterns. The locations of recirculation zones and short-circuits can be identified with this measuring technique, resulting in a better understanding of the local flow features. The results are however difficult to quantify, so this technique can only be used for validation in a qualitative manner.
- Laser Doppler anemometry measures the instantaneous velocities at a single point in the system, resulting in an accurate measurement of the mean velocities as well as the turbulent properties (Reynolds stresses) of the flow. This technique provides valuable information for validation purposes, since it precisely points out where the inaccuracies of the CFD model may occur.
- Lagrangian actinometry measures the dose distribution in a UV system. By using this technique, the combined effect of the hydraulics and the radiation distribution is measured, characterising the performance of a UV system precisely. This technique is therefore useful for validation, although inaccuracies might arise due to the convolution of the experimental data to obtain the dose distribution.
- Biodosimetry measures the inactivation of micro-organisms. It is used for validation, since it provides the actual disinfection performance, which is the final result we want to predict. Due to the high inactivation levels, the measurements are sensitive to the local hydraulics.
- Measurements of degradation of chemical substances provide the actual oxidation performance of the system. It is also useful for validation of the CFD model, but mainly for the mean dose, since the extent of degradation is usually low. Hence, the sensitivities with respect to the local flow fields are not clarified by this measurement technique.

Residence time distributions were measured for a full-scale ozone contactor that served as a validation of the CFD model, showing a good agreement between the model and experiment (Chapter 4). Dye visualisation and laser Doppler anemometry were used to measure the flow around a single cross-flow UV lamp, in order to investigate the wake formation behind a UV lamp, and to validate the flow fields for both modelling techniques (Chapter 5). The LES model approached the measured velocity profiles most closely, whereas the k-ε model under predicted the wake downstream of the lamp. All of the above mentioned methods were used to validate a bench-scale UV reactor (Chapter 6). The LES model predicted the flow fields and residence time distributions correctly, whereas the k-ε model under predicted the formation of recirculations. The dose distribution measured by Lagrangian actinometry showed some variabilities, but in general the shape of the dose distribution was predicted well by both models. The microbial inactivation was under predicted by both CFD models, which was related to the variability in the dose-response behaviour of the micro-organisms and uncertainty in lamp power and UV irradiation field in the models. The degradation of chemicals was predicted accurately by both models.

7.1.4 Use of CFD models in the water industry

Due to the advances in computer engineering CFD models become more user friendly and more widely available, so that an increasing number of people can use these models to simulate processes occurring in all kinds of installations. However, a CFD model should not be regarded as off-the-shelf software, a thorough mathematical and physical background is required for a proper interpretation of the results. Also, the limitations of the model should be acknowledged, one must be aware of the numerical inaccuracies and the various assumptions made in the model, for instance to model the turbulence. One of the issues addressed in this thesis was an evaluation of the effect of different turbulent models on the prediction of various quantities (residence time, dose, microbial inactivation, degradation) in UV systems. Using this experience, a scheme is developed (Figure 7.2) that could assist CFD users in deciding at which detail the turbulence needs to be resolved: solving either a RANS model or a large-eddy simulation. The large-eddy simulation accounts for the large-scale turbulent motions varying in time and space, resulting in the more accurate solution of the hydrodynamics. The choice for the model depends on fundamental questions, such as the required information, complexity of the flow domain and to what extent recirculations can develop, and practical considerations, such as the size of the computational domain and the available time for computing. By answering each question, starting from the top, this scheme will give an advice which model to use. Once an answer in the most right compartment is found, we will advice to use a RANS model. Depending on the size of the domain and available time of computation, the resolution of the mesh can be adjusted to obtain the most accurate result for each type of model. Using this scheme does not give the certainty that the results of the model will be correct - the complex fluid dynamics can of course not be squeezed in such a simple scheme, but shows which model is most probably the best for the user's specific application with the present day computers. Experimental validation is therefore still highly recommended.

7.2 SYSTEM DESIGN CONSIDERATIONS

The hydraulic processes that are responsible for a suboptimal performance of treatment installations are in general related to short-circuiting and recirculation zones. The combination of short-circuits causing short residence times and recirculation zones causing long residence times results in a suboptimal hydraulic behaviour of the system. This is well illustrated by the concept of axial dispersion (Taylor, 1954), where differences in velocity profile increase dispersion, and a cross-stream diffusion process reduces the dispersion by mixing water elements with different velocities over the cross-section. Hydraulic improvements are mainly focussed on stimulating the mixing in cross-stream direction, so that the short-circuits and dead zones are mixed with each other, and on streamlining the system, so that the dead zones are reduced resulting in a smoother velocity profile. The result is a narrowing of the residence time distributions. However, improvement of the residence time distributions does not necessarily lead to an improvement of the disinfection (or oxidation) performance, since the spatial distribution of disinfectant (or oxidiser) is not incorporated in the residence time distribution. This is particularly important for UV systems, where short residence times can be compensated by high irradiance intensities, and long residence times can be compensated by low irradiance intensities. So, the dose distribution needs to be considered for these systems to characterise the performance. For ozone systems this effect is less pronounced, because the ozone concentration also depends on residence times.

The improvements are usually established by geometric modifications. Baffles are common modifications that increase the mixing in cross-stream direction, which may however contradict with the demand of streamlining the flow, since baffles themselves cause a recirculation zone. In UV systems, baffles are also used to force the main flow to come close to the lamp, allowing a recirculation at a less critical region - a region where the UV intensities are low. By adding baffles a little downstream in a single cross-flow UV lamp system we showed that, although the residence time distributions became wider, the dose distributions became smaller, resulting in a better disinfection performance. The short residence times could be compensated with high irradiances, because these short-circuits came close to the lamps. However, baffles will not always have the desired effect, for example if they cause recirculations at the wrong locations (close to the lamp).

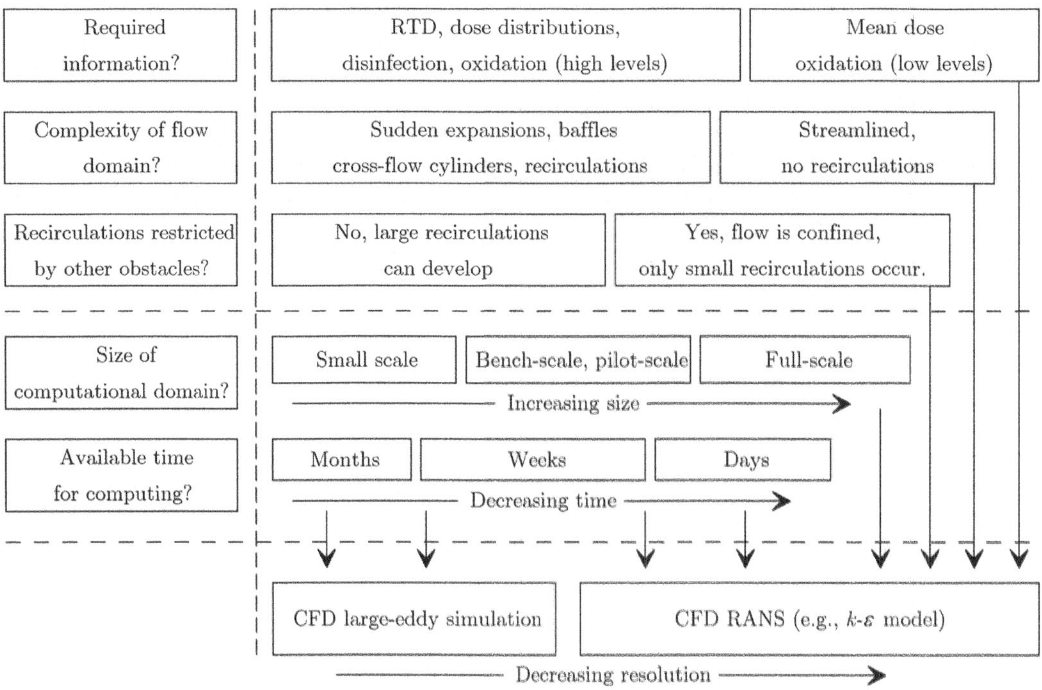

Figure 7.2 This scheme advices a CFD user on which model to use for different applications. Starting from the top, once an answer to the question is found in the most right compartment, the advice is to use a RANS model.

To quantify the effect of hydraulic optimisations, we used a simple analytical model to represent the dose distribution. It was shown that a shifted exponential probability density function (eq. 2.32) was a good estimate for the dose distribution that occurs in treatment installations, and the factor D_{10}/D_{mean} was a good characterisation of this dose distribution. The actual gain of a hydraulic optimisation mainly depends on the reduction in dose distribution and the removal target. The effect of reducing the dose distribution is more pronounced at a higher removal target. Using the simple analytical model the effect of hydraulic optimisations are summarised in Figure 7.3. In the left panel different dose distributions are shown, and in the right panel the consumption as a function of dose distribution is shown for different removal targets. This consumption can be energy, chemicals, etc., all the properties that are proportional to the mean dose. It is shown as an additional consumption required with respect to the ideal situation, which represents the theoretical maximum with a dose distribution of zero. The grey box in Figure 7.3 shows the range of dose distribution of current installations (ozone contactors or UV systems), calculated by the CFD model. Depending on the removal target, which is around 3 log for disinfection and less than 1 log for oxidation, and on the efficacy of current installations, hydraulic optimisations can establish a reduction in consumption of 10% to 60%. This reduction is worth the effort, since power consumptions of UV reactors are high, and reduction of ozone dosage in ozone systems reduces the formation of by-products.

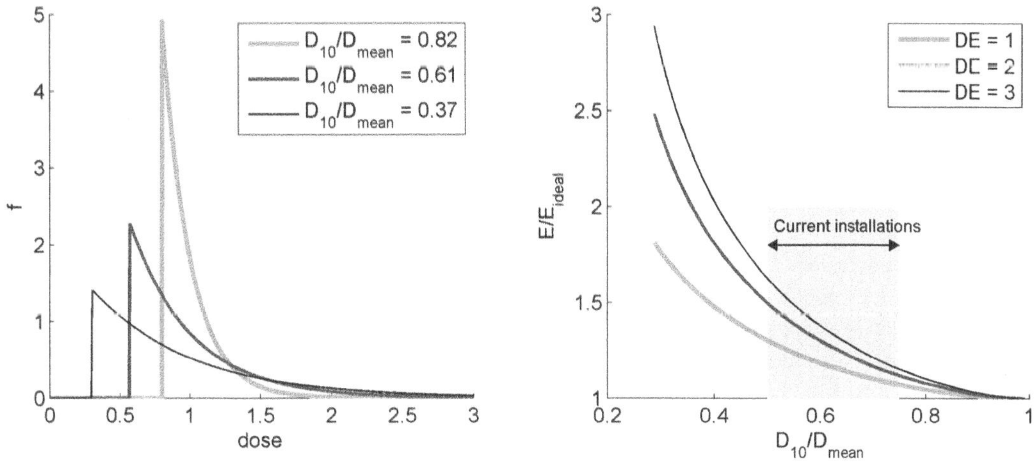

Figure 7.3 Reduction in consumption (e.g., energy, chemicals) by optimising the hydraulics. The left panel shows the dose distributions approximated by a shifted exponential distribution, and the right panel shows the consumption (normalised by the consumption for the ideal situation) as a function of dose distribution (characterised by D_{10}/D_{mean}).

7.2.1 Ozone systems

In ozone contactors, modifications were proposed for the Leiduin ozone contactor, such as additional horizontal baffles or turning vanes. The CFD model showed that these measures reduced the short-circuits and large recirculations zones, so that the disinfection performance was increased significantly, whereas the mean CT value remained the same. As a result, the bromate production, which is proportional to the mean CT value, will not be increased by the hydraulic improvements. The gain in disinfection performance depends on the kinetics, represented by the slow ozone decay coefficient and the microbial inactivation rate. Optimising the hydraulics has little effect at small decay coefficients or for micro-organisms that are very resistant to ozone (small inactivation rate constants). But in the range of required log removal for micro-organisms encountered in practice, improving the hydraulics has a significant effect on the disinfection performance.

7.2.2 UV systems

The disinfection performance, oxidation performance and energy consumption of a wide range of UV reactors was assessed. Although the input parameters such as lamp power, flow rate and water transmittance were the same, the dose distributions differed greatly for the various UV systems. Hence, the differences between the disinfection levels predicted by the CFD model were large. Since hydraulic energy losses are small with respect to the lamp energy, increasing the mixing in a UV reactor will reduce the total energy consumption. An annular system, designed as a static mixer with a lamp in the middle, showed a very narrow dose distribution, which comes close to the perfect hydraulic system ($D_{10}/D_{\text{mean}} > 0.8$, which is within 10% of the energy consumption of the ideal system, Figure 7.3). The efficacy of UV systems can be improved by taking measures that increase the mean dose and/or narrow the dose distribution:

- Enhance the mixing inside the system to ensure that all the micropollutants come close to the lamps. This can be achieved by placing baffles or by forcing a helical flow around the lamp. However, when considering baffles to improve UV reactors, the position of the baffles needs to be chosen carefully. For a single cross-flow UV lamp system, placing the baffles a little upstream of the centre of the lamp resulted in a less efficient disinfection than for a case without a baffle. Moving the baffle a little downstream of the centre of the lamp resulted in an increase in disinfection.
- Place mirrors at the outer walls, especially in the case of an annular system, which leads to a significant increase in the mean dose and therefore in disinfection and oxidation levels.
- Reduce the wake size downstream of the UV lamp. Measurements were conducted to investigate the effect of hydraulic changes that reduce the wake size downstream of UV lamps. Placing a grid upstream of the lamp resulted in a shorter wake length, and a higher disinfection.
- Place a number of reactors in series with a static mixer in between the reactors. Due to the mixing in between the reactors, the dose distribution narrows after each reactor, so that a number of bad designed reactors in series will eventually result in a narrow dose distribution.

Other design considerations are the orientation, size and shape of the lamps. From the simulations, there was a slight preference found for an annular system but no clear preference for the size of the lamp. Experiments showed that increasing the roughness of the lamp surface did not reduce the wake size. Also, streamlining the lamp resulted in a disappearance of the wake in case the Reynolds number was high enough (above $Re = 24\ 000$), the disinfection was however not improved. Although some general design rules can be formulated, due to the complexity of the flow fields and large spatial differences in UV irradiance, CFD is an essential tool to optimise the design of an existing UV reactor.

Optimal distance from lamp to outer wall

For the design of UV reactors the distance of the lamps to the outer wall is an important parameter: it must not be too large to prevent areas with small doses and it must not be too small to avoid the loss of radiant energy through the outer walls. A simple analytical model was developed to calculate the optimal distance from the quartz sleeve of the lamp to the outer wall (§2.7.2). This model was based upon an annular system containing one lamp in the middle, but could also be used as a rule of thumb for cross-flow reactors. The optimal distance was a function of the water transmittance, equal to $-0.01/\ln(T_w)$, and is shown in Table 7.1 for different transmittances.

Table 7.1 Optimal distance from the lamp to the outer wall of the reactor.

Transmittance (%)	60	65	70	75	80	85	90	95
Optimal distance (cm)	2.0	2.3	2.8	3.5	4.5	6.2	9.5	19

7.3 OUTLOOK FOR CFD MODELLING IN DISINFECTION/OXIDATION PROCESSES

The ongoing development of computer technology allows us to increase the mesh resolution of numerical models further and to resolve more and more details of the turbulence. Moore's law states that roughly every two year the computer power is doubled (Moore, 2005), so that in the future perhaps all the turbulent scales can be resolved for full-scale installations. In Table 7.2 an overview is given of the computation times required to model a bench-scale UV reactor and a full-scale UV reactor. For the bench-scale UV reactor, the RANS and LES computations described in this thesis showed feasible calculation times (within days or weeks). Using a direct numerical simulation (DNS) that resolves all the turbulent scales would however require 13 years of computation on a present day workstation (number of mesh elements $\sim \frac{L_{\text{box}}}{D} Re^{9/4}$, where L_{box} is the size of the domain, and number of time steps $\sim \frac{T}{D/u_m} Re^{3/4}$, where T is the total simulation time, Davidson (2004)). For full-scale installations the dimensions and velocities are larger than for the bench-scale installations. Consequently, the Reynolds numbers become larger, but also the smallest turbulent scales become smaller. Therefore, computational times increase dramatically, so that LES and DNS computations are out of reach for present day computers. We have made a rough estimate how long it will take before a simulation is finished within a day, for which we have to wait at least 60 years for a DNS of the full-scale installation.* In 30 years, perhaps a LES of the full-scale or a DNS of the bench-scale is within reach.

Table 7.2 Comparison of computing time for different CFD models.

	Bench-scale	Full-scale
Reynolds number	$1.4{\cdot}10^4$	$1{\cdot}10^6$
Computing time*		
RANS	0.5 day	5 days
LES	3 weeks	57 years
DNS	13 years	$5{\cdot}10^6$ years
Years before computation is completed in 1 day		
RANS	0	5
LES	9	29
DNS	25	62

*Using a large workstation: 4 CPUs of 2 GHz, 32 GB memory, around 2 Gigaflop. As a reference for the DNS, a computation of $Re = 500$, $L_{box} = 10D$ and $T = 5D/u_m$ would take 9 days on a 1 Gigaflop CPU (Davidson, 2004). For our reactors, we use $L_{box} = 2D$ and $T = 30D/u_m$.

Another constraint on computational modelling of drinking water treatment systems is reaction kinetics. Calculation of all the reactions occurring in the installation (e.g., oxidation by ozone or UV/H_2O_2) would require a huge computational effort, especially in the case of high reaction rates, such as the reaction of $OH\bullet$ with all kinds of substances. Also, a thorough analysis of the water quality is then required to obtain

* Assuming that Moore's law, which is purely empirical evidence that the computer power is doubled roughly every two years, still holds over 60 years might be questionable, especially because transistors would eventually meet the limitations of atomic levels. Moreover, electrical power consumptions will increase dramatically. However, nobody expected that it would still be valid today when Moore proposed his law 40 years ago. Or to cite Moore: '*Materials are made of atoms, and we're getting suspiciously close to some of the atomic dimensions with these new structures, but I'm sure we'll find ways to squeeze even further than we think we presently can.*' (Moore, 2005).

the initial concentrations of all the substances in the water. For multiphase flow systems (such as the ozone bubble column), additional computational resources are required to predict the multiphase hydrodynamics, interfacial mass transfer and chemical reactions. Before a major breakthrough in computer technology is established, modelling a large amount of chemical reactions requires a decoupling of the chemical model from the hydrodynamic model, so that the chemistry is resolved in a coarser compartmental model using the information from the hydrodynamics (Rigopoulos and Jones, 2003).

Experimental validation by laser Doppler anemometry or particle image velocimetry is useful to reveal the inaccuracies of the hydrodynamics in UV systems. In the future, these inaccuracies could be reduced by using the more complex models. However, another uncertainty in the modelling of UV reactors is the distribution of UV radiation. These uncertainties are caused by inhomogeneities of the lamp output over the arc length of the lamp, ageing of the lamps, fouling of the quartz sleeves and the temperature dependence of the lamp output. Experimental validation techniques of CFD models for UV reactors should therefore also account for the spatial distribution of UV radiation in the reactor. Due to the variations of lamp output in time, the challenge for a proper validation would be to measure the UV radiation distribution at the same instant as the microbial inactivation. Other issues that require attendance in the modelling of UV systems are:

- Ambient particles could influence the distribution of UV radiation by shadowing and scattering of radiation, mainly an issue for low-quality water.
- Temperature variations due to the heating of water by UV lamps with high UVC output (medium pressure lamps). The increase in water temperature close to the lamp results in density-driven flows, which requires heat transfer modelling.
- Lamp fouling influences the UV radiation field and may also influence the hydrodynamics due to a higher roughness of the quartz sleeve (especially of interest for low-quality water).

In conclusion: in the coming decade, the water industry is still obliged to use simplified models that do not resolve all the details of the turbulent (multiphase) flow and chemical reactions. Therefore, as demonstrated in this thesis, one should be aware of the limitations of the CFD models, and validate the models by experiments.

References

Achenbach, E. (1971). Influence of surface roughness on the cross-flow around a circular cylinder. *Journal of Fluid Mechanics*, 46(2):321–335.

Achenbach, E. and Heinecke, E. (1981). On vortex shedding from smooth and rough cylinders in the range of Reynolds numbers $6*10^3$ to $5*10^6$. *Journal of Fluid Mechanics*, 109:239–251.

Adams, E. W. and Rodi, W. (1990). Modeling flow and mixing in sedimentation tanks. *Journal of Hydraulic Engineering*, 116(7):895–913.

Alpert, S. M., Knappe, D., and Ducoste, J. (2007). The use of computational fluid dynamics (CFD) to model UV-initiated advanced oxidation processes. In *2007 World Congress on Ozone and Ultraviolet Technologies*, Los Angeles.

Aris, R. (1959). Notes on the diffusion-type model for longitudinal mixing in flow. *Chemical Engineering Science*, 9:266–267.

AWWA (1991). *Guidance manual for compliance with the filtration and disinfection requirements for public water systems using surface water sources*. Denver, CO.

Baldyga, J. and Orciuch, W. (2001). Barium sulphate precipitation in a pipe - an experimental study and CFD modelling. *Chemical Engineering Science*, 56:2435–2444.

Blatchley III, E. R., Do-Quang, Z., Janex, M. L., and Laîné, J. M. (1998). Process modeling of ultraviolet disinfection. *Water Science and Technology*, 38(6):63–69.

Blatchley III, E. R., Shen, C., Naunovic, Z., Lin, L. S., Lyn, D. A., Robinson, J. P., Ragheb, K., Grgori, G., Bergstrom, D. E., Fang, S., Guan, Y., Jennings, K., and Gunaratna, N. (2006). Dyed microspheres for quantification of UV dose distributions: Photochemical reactor characterization by Lagrangian actinometry. *Journal of Environmental Engineering*, 132(11):1390–1403.

Blatchley III, E. R., Shen, C., Scheible, O. K., Robinson, J. P., Ragheb, K., Bergstrom, D. E., and Rokjer, D. (2008). Validation of large-scale, monochromatic UV disinfection systems for drinking water using dyed microspheres. *Water Research*, 42(3):677–688.

Bolton, J. R. (2000). Calculation of ultraviolet fluence rate distributions in an annular reactor: significance of refraction and reflection. *Water Research*, 34(13):3315–3324.

Bolton, J. R., Dussert, B., Bukhari, Z., Hargy, T., and Clancy, J. L. (1998). Inactivation of *Cryptosporidium parvum* by medium-pressure ultraviolet light in finished drinking water. In *Proceedings AWWA Annual Conference*, volume Vol. A, pages 389–403, Dallas TX.

Bolton, J. R. and Linden, K. G. (2003). Standardization of methods for fluence (UV dose) determination in bench-scale UV experiments. *Journal of Environmental Engineering*, 129(3):209–215.

Bolton, J. R. and Stefan, M. I. (2002). Fundamental photochemical approach to the concepts of fluence (UV dose) and electrical energy efficiency in photochemical degradation reactions. *Research on Chemical Intermediates*, 28(7-9):857–870.

Breuer, M. (2000). A challenging test case for large eddy simulation: high Reynolds number circular cylinder flow. *International Journal of Heat and Fluid Flow*, 21(5):648–654.

Brown, R. (1829). Additional remarks of active molecules. *Philosophical Magazine NS*, 6:161–166.

Charles, W. M. (2007). *Transport modelling in coastal waters using stochastic differential equations*. PhD thesis, Delft University of Technology.

Chick, H. (1908). An investigation of the laws of disinfection. *Journal of Hygiene*, 8:92–158.

Chiu, K., Lyn, D. A., Savoye, P., and Blatchley III, E. R. (1999). Integrated UV disinfection model based on particle traking. *Journal of Environmental Engineering*, 125(1):7–16.

Cockx, A., Do-Quang, Z., Audic, J. M., Liné, A., and Roustan, M. (2001). Global and local mass transfer coefficients in waste water treatment process by computational fluid dynamics. *Chemical Engineering and Processing*, 40:187–194.

Cockx, A., Do-Quang, Z., Liné, A., and Roustan, M. (1999). Use of computational fluid dynamics for simulating hydrodynamics and mass transfer in industrial ozonation towers. *Chemical Engineering Science*, 54(21):5085–5090.

Cotruvo, J., Bull, R., Cummings, B., Fisher, J., Guo, Z., Nam Ong, C., Quinones, O., Snyder, S., Keith, J., Gordon, G., and Pacey, G. (2009). Bromate pre-systemic detoxification metabolism research progress. In *19th Ozone World Congress & Exhibition*, Tokyo, Japan.

Courant, R., Friedrichs, K., and Lewy, H. (1928). Über die partiellen Differenzengleichungen der Mathematischen Physik. *Mathematische Annalen*, 100(1):32–74.

Craig, K., De Traversay, C., Bowen, B., Essemiani, K., Levecq, C., and Naylor, R. (2002). Hydraulic study and optimisation of water treatment processes using numerical simulation. *Water Science and Technology: Water Supply*, 2(5-6):135–142.

Craik, S. A., Finch, G. R., Bolton, J. R., and Belosevic, M. (2000). Inactivation of *Giardia muris* cysts using medium-pressure ultraviolet radiation in filtered drinking water. *Water Research*, 34(18):4325–4332.

Craik, S. A., Weldon, D., Finch, G. R., Bolton, J. R., and Belosevic, M. (2001). Inactivation of *Cryptosporidium parvum* oocysts using medium- and low-pressure ultraviolet radiation. *Water Research*, 35(6):1387–1398.

Craik, S. A. (2005). Effect of micro-mixing conditions on predictions of *Cryptosporidium* inactivation in an ozone contactor. *Ozone: Science & Engineering*, 27(6):487–494.

Davidson, P. A. (2004). *Turbulence: An introduction for scientists and engineers*. Oxford University Press, Oxford.

de Laat, J., Gallard, H., Ancelin, S., and Legube, B. (1999). Comparative study of the oxidation of atrazine and acetone by H_2O_2/UV, Fe(III)/UV, Fe(III)/H_2O_2/UV and Fe(II) or Fe(III)/H_2O_2. *Chemosphere*, 39(15):2693–2706.

de Moel, P. J., Verberk, J. Q. J. C., and van Dijk, J. C. (2004). *Drinkwater - principes en praktijk*. Den Haag, NL.

Dixon, A. G. and Nijemeisland, M. (2001). CFD as a design tool for fixed-bed reactors. *Industrial & Engineering Chemistry Research*, 40(23):5246–5254.

Ducoste, J. J., Liu, D., and Linden, K. (2005). Alternative approaches to modeling fluence distribution and microbial inactivation in ultraviolet reactors: Lagrangian versus Eulerian. *Journal of Environmental Engineering*, 131(10):1393–1403.

Ducoste, J., Carlson, K., and Bellamy, W. (2001). The integrated disinfection design framework approach to reactor hydraulics characterization. *Journal of Water Supply: Research and Technology – AQUA*, 50(4):245–261.

Dunsbergen, D. W. (1994). *Particle models for transport in three-dimensional shallow water flow*. PhD thesis, Delft University of Technology.

Einstein, A. (1905). Über die von der molekularkinetischen Theorie der Wärme geforderte Bewegung von in ruhenden Flüssigkeiten suspendierten Teilchen. *Annalen der Physik*, 322(8):549–560.

Elder, J. W. (1959). The dispersion of marked fluid in turbulent shear flow. *Journal of Fluid Mechanics*, 5(4):544–560.

Elíasson, J., Kjaran, S. P., Holm, S. L., Gudmundsson, M. T., and Larsen, G. (2007). Large hazardous floods as translatory waves. *Environmental Modelling & Software*, 22(10):1392–1399.

Ghidossi, R., Veyret, D., and Moulin, P. (2006). Computational fluid dynamics applied to membranes: State of the art and opportunities. *Chemical Engineering and Processing*, 45(6):437–454.

Greene, D. J., Haas, C. N., and Farouk, B. (2006). Computational fluid dynamics analysis of the effects of reactor configuration on disinfection efficiency. *Water Environment Research*, 78(9):909–919.

Ham, D. A., Pietrzak, J., and Stelling, G. S. (2006). A streamline tracking algorithm for semi-Lagrangian advection schemes based on the analytic integration of the velocity field. *Journal of Computational and Applied Mathematics*, 192(1):168–174.

Hanjalić, K. and Jakirlić, S. (1998). Contribution towards the second-moment closure modelling of separating turbulent flows. *Computers and Fluids*, 27(2):137–156.

Hijnen, W. A. M., Beerendonk, E. F., and Medema, G. J. (2006). Inactivation credit of UV radiation for viruses, bacteria and protozoan (oo)cysts in water: A review. *Water Research*, 40(1):3–22.

Hilgenstock, A. and Ernst, R. (1996). Analysis of installation effects by means of computational fluid dynamics - CFD vs experiments? *Flow Measurement and Instrumentation*, 7(3-4):161–171.

Huang, T. H., Brouckaert, C. J., Pryor, M., and Buckley, C. A. (2004). Application of computational fluid dynamics modelling to an ozone contactor. *Water SA*, 30(1):51–56.

ISO 10705-1 (1995). *Water quality - Detection and enumeration of bacteriophages - Part 1: Enumeration of F-specific RNA bacteriophages*. International Organization for Standardization.

Kappelhof, J., van Breukelen, L., Dullemont, Y. J., Nienhuis, P., and van der Aa, R. (2007). Desinfectiescenario's Leiduin. Technical report, Internal report Waternet, 07.004572.

Kloeden, P. E. and Platen, E. (1992). *Numerical solution of stochastic differential equations*. Applications of mathematics vol. 23. Springer, Berlin.

Kovalets, I. V., Andronopoulos, S., Venetsanos, A. G., and Bartzis, J. G. (2008). Optimization of the numerical algorithms of the ADREA-I mesoscale prognostic meteorological model for real-time applications. *Environmental Modelling & Software*, 23(1):96–108.

Kruithof, J. C., Kamp, P. C., and Martijn, B. J. (2007). UV/H_2O_2 treatment: A practical solution for organic contaminant control and primary disinfection. *Ozone: Science & Engineering*, 29(4):273 – 280.

KWR (2004). Inactivation of anaerobic spores (SSRC) by UV in natural water. Technical report, BTO 2003.052.

KWR (2009). Omzetting van prioritaire stoffen met UV/H_2O_2 oxidatie. Technical report, BTO 2009.036.

Labeur, R. J. and Wells, G. N. (2007). A Galerkin interface stabilisation method for the advection-diffusion and incompressible Navier-Stokes equations. *Computer Methods in Applied Mechanics and Engineering*, 196(49-52):4985–5000.

Launder, B. E. and Spalding, D. B. (1974). The numerical computation of turbulent flows. *Computer Methods in Applied Mechanics and Engineering*, 3(2):269–289.

Levenspiel, O. and Smith, W. K. (1957). Notes on the diffusion-type model for the longitudinal mixing of fluids in flow. *Chemical Engineering Science*, 6(4-5):227–235.

Levenspiel, O. (1972). *Chemical Reactor Engineering*. John Wiley, New York.

Li, J., Zhang, J., Miao, J., Ma, J., and Dong, W. (2006). Application of computational fluid dynamics (CFD) to ozone contactor optimization. *Water Science and Technology: Water Supply*, 6(4):9–16.

Liu, D., Ducoste, J. J., Jin, S., and Linden, K. (2004). Evaluation of alternative fluence rate distribution models. *Journal of Water Supply: Research and Technology – AQUA*, 53(6):391–408.

Liu, D., Wu, C., Linden, K., and Ducoste, J. (2007). Numerical simulation of UV disinfection reactors: Evaluation of alternative turbulence models. *Applied Mathematical Modelling*, 31(9):1753–1769.

Liu, Y. and Ducoste, J. (2006). Numerical simulation of chloramines formation in turbulent flow using a multi-fluid micromixing model. *Environmental Modelling & Software*, 21(8):1198–1213.

Lübcke, H., Schmidt, S., Rung, T., and Thiele, F. (2001). Comparison of LES and RANS in bluff-body flows. *Journal of Wind Engineering and Industrial Aerodynamics*, 89(14-15):1471–1485.

Lyn, D. A. and Blatchey III, E. R. (2005). Numerical computational fluid dynamics-based models of ultraviolet disinfection channels. *Journal of Environmental Engineering*, 131(6):838–849.

Lyn, D. A., Chiu, K., and Blatchley III, E. R. (1999). Numerical modeling of flow and disinfection in UV disinfection channels. *Journal of Environmental Engineering*, 125(1):17–26.

Martin, A. D. (2000). Interpretation of residence time distribution data. *Chemical Engineering Science*, 55(23):5907–5917.

Martin-Dominguez, A., Tzatchkov, V. G., Martin-Dominguez, I. R., and Lawler, D. F. (2005). An enhanced tanks-in-series model for interpretation of tracer tests. *Journal of Water Supply: Research and Technology - AQUA*, 54(7):435–448.

Modi, S. (1994). Interaction of aromatic donor molecules with horseradish peroxidase: identification of the binding site and role of heme iron in the binding and activity. *BioMetals*, 8:218–222.

Moeng, C. (1984). A large-eddy-simulation model for the study of planetary boundary-layer turbulence. *Journal of the Atmospheric Sciences*, 41(13):2052–2062.

Moore, G. (2005). Excerpts from a conversation with Gordon Moore: Moore's law. *Intel Corporation*.

Munoz, A., Craik, S., and Kresta, S. (2007). Computational fluid dynamics for predicting performance of ultraviolet disinfection - sensitivity to particle tracking inputs. *Journal of Environmental Engineering and Science*, 6(3):285–301.

Nick, K., Schoeler, H., Mark, G., Soylemez, T., Akhlaq, M., Schuchmann, H.-P., and von Sonntag, C. (1992). Degradation of some triazine herbicides by UV radiation such as used in the UV disinfection of drinking water. *Journal of Water Supply: Research and Technology - AQUA*, 41(2):82–87.

Niemann, H. J. and Hölscher, N. (1990). A review of recent experiments on the flow past circular cylinders. *Journal of Wind Engineering and Industrial Aerodynamics*, 33(1-2):197–209.

ÖNORM (2001). Requirements and testing (low pressure mercury lamp plants), Austrian national standard ÖNORM M 5873-1 (batch 2007).

Pan, H. and Orava, M. (2007). Performance evaluation of the UV disinfection reactors by CFD and fluence simulations using a concept of disinfection efficiency. *Journal of Water Supply: Research and Technology – AQUA*, 56(3):181–189.

Parsons, S. (2004). *Advanced Oxidation Processes for Water and Wastewater Treatment*. IWA Publising, London, England.

Rakness, K. L., Najm, I., Elovitz, M., Rexing, D., and Via, S. (2005). *Cryptosporidium* log-inactivation with ozone using effluent CT10, geometric mean CT10, extended integrated CT10 and extended CSTR calculations. *Ozone: Science & Engineering*, 27(5):335–350.

Rennecker, J. L., Mariñas, B. J., Owens, J. H., and Rice, E. W. (1999). Inactivation of *Cryptosporidium parvum* oocysts with ozone. *Water Research*, 33(11):2481–2488.

Rietveld, L. C., van Dijk, J. C., van der Helm, A. W. C., van Schagen, K., and van der Aa, R. (2008). Integrated simulation of drinking water treatment. *Journal of Water Supply: Research and Technology – AQUA*, 57(3):133–141.

Rietveld, L. C. (2005). *Improving operation of drinking water treatment through modelling*. PhD thesis, Delft University of Technology.

Rigopoulos, S. and Jones, A. (2003). A hybrid CFD–reaction engineering framework for multiphase reactor modelling: basic concept and application to bubble column reactors. *Chemical Engineering Science*, 58(14):3077–3089.

Risken, H. (1984). *The Fokker-Planck Equation*. Springer series in synergetics 18. Springer, Berlin.

Rodi, W. (1997). Comparison of LES and RANS calculations of the flow around bluff bodies. *Journal of Wind Engineering and Industrial Aerodynamics*, 69-71:55–75.

Rokjer, D., Matthew, V., Keesler, D., Boryskowsky, M., and Ducoste, J. (2003). Medium pressure UV reactor models for validation purposes. In *Proceedings of the American Water Works Association Water Quality Technology Conference*, Philadelphia.

Ruck, B. (1987). Laser Doppler anemometry - a non-intrusive optical measuring technique for fluid velocity. *Particle & Particle Systems Characterization*, 4(1-4):26–37.

Schwager, A. and Boller, M. (1997). Transport phenomena in intermittent filters. *Water Science and Technology*, 35(6):13–20.

Sharpless, C. M. and Linden, K. G. (2003). Experimental and model comparisons of low- and medium-pressure Hg lamps for the direct and H_2O_2 assisted UV photodegradation of N-nitrosodimethylamine in simulated drinking water. *Environmental Science and Technology*, 37(9):1933–1940.

Smeets, P. W. M. H. (2008). *Stochastic modelling of drinking water treatment in quantitative microbial risk assessment*. PhD thesis, Delft University of Technology.

Smeets, P. W. M. H., van der Helm, A. W. C., Dullemont, Y. J., Rietveld, L. C., van Dijk, J. C., and Medema, G. J. (2006). Inactivation of *Escherichia coli* by ozone under bench-scale plug flow and full-scale hydraulic conditions. *Water Research*, 40(17):3239–3248.

Solazzo, E., Cai, X., and Vardoulakis, S. (2009). Improved parameterisation for the numerical modelling of air pollution within an urban street canyon. *Environmental Modelling & Software*, 24(3):381–388.

Sommer, R., Cabaj, A., and Haider, T. (1996). Microbicidal effect of reflected UV radiation in devices for water disinfection. *Water Science and Technology*, 34(7-8):173–177.

Sommer, R., Cabaj, A., Hirschmann, G., and Haider, T. (2008). Disinfection of drinking water by UV irradiation: Basic principles - specific requirements - international implementations. *Ozone: Science & Engineering*, 30(1):43 – 48.

Sozzi, A. and Taghipour, F. (2007). The importance of hydrodynamics in UV advanced oxidation reactors. *Water Science and Technology*, 55(12):53–58.

Sozzi, D. A. and Taghipour, F. (2006). UV reactor performance modeling by Eulerian and Lagrangian methods. *Environmental Science and Technology*, 40(5):1609–1615.

Stijnen, J. (2002). *Numerical methods for stochastic environmental models*. PhD thesis, Delft University of Technology.

Taylor, G. (1954). The dispersion of matter in turbulent flow through a pipe. *Proceedings of the Royal Society of London, Series A, Mathematical and Physical Sciences*, 223(1155):446–468.

USEPA (1991). *Guidance manual for compliance with the filtration and disinfection requirements for public water systems using surface water sources*. U.S. Environmental Protection Agency.

USEPA (2006a). *Long term 2 enhanced surface water treatment rule and draft toolbox guidance manual (LT2ESWTR)*. U.S. Environmental Protection Agency.

USEPA (2006b). *Ultraviolet disinfection guidance manual for the final long term 2 enhanced surface water treatment rule*. U.S. Environmental Protection Agency.

van Balen, W., Uijttewaal, W. S. J., and Blanckaert, K. (2009). Large-eddy simulation of a mildly curved open-channel flow. *Journal of Fluid Mechanics*, 630:413–442.

van der Helm, A. W. C. (2007). *Integrated modeling of ozonation for optimization of drinking water treatment*. PhD thesis, Delft University of Technology.

van der Helm, A. W. C., Smeets, P. W. M. H., Baars, E. T., Rietveld, L. C., and van Dijk, J. C. (2007). Modeling of ozonation for dissolved ozone dosing. *Ozone: Science & Engineering*, 29(5):379–389.

van der Meer, W. G. J. (2003). *Mathematical modelling of NF and RO membrane filtration plants and modules*. PhD thesis, Delft University of Technology.

van der Walt, J. J. (2002). *The modelling of water treatment process tanks*. PhD thesis, Rand Afrikaans University.

van Kämpen, N. G. (1981). Itô versus stratonovich. *Journal of Statistical Physics*, 24(1):175–187.

van Schagen, K. (2009). *Model-based control of drinking-water treatment plants*. PhD thesis, Delft University of Technology.

Verberk, J. Q. J. C. (2005). *Application of air in membrane filtration*. PhD thesis, Delft University of Technology.

Verliefde, A. (2008). *Rejection of organic micropollutants by high pressure membranes (NF/RO)*. PhD thesis, Delft University of Technology.

von Gunten, U. (2003a). Ozonation of drinking water: Part I. oxidation kinetics and product formation. *Water Research*, 37(7):1443–1467.

von Gunten, U. (2003b). Ozonation of drinking water: Part II. disinfection and by-product formation in presence of bromide, iodide or chlorine. *Water Research*, 37(7):1469–1487.

Vreeburg, J. H. G. (2007). *Discolouration in drinking water systems: a particular approach*. PhD thesis, Delft University of Technology.

VROM (2001). *Waterleidingbesluit (drinking water decree)*. Ministry of Housing, Physical Planning and the Environment, The Hague, The Netherlands.

Watson, H. E. (1908). A note on the variation of the rate of disinfection with change in the concentration of the disinfectant. *Journal of Hygiene*, 8:536–542.

Wille, R. and Fernholz, H. (1965). Report on the first European mechanics colloquium, on the Coanda effect. *Journal of Fluid Mechanics*, 23(4):801–819.

Wright, N. G. and Hargreaves, D. M. (2001). The use of CFD in the evaluation of UV treatment systems. *Journal of Hydroinformatics*, 3(2):59–70.

Zhang, J., Huck, P. M., Anderson, W. B., and Stubley, G. D. (2007). A computational fluid dynamics based integrated disinfection design approach for improvement of full-scale ozone contactor performance. *Ozone: Science & Engineering*, 29(6):451 – 460.

Zhou, H. and Smith, D. W. (1994). Kinetics of ozone disinfection in completely mixed system. *Journal of Environmental Engineering*, 120(4):841–858.

Zwietering, T. N. (1959). The degree of mixing in continuous flow systems. *Chemical Engineering Science*, 11(1):1–15.

Appendix A

Modelling approaches

STANDARD *K*-ε MODEL

In the k-ε model, the time-varying turbulent motions are captured in the averaged variables for the turbulent kinetic energy (k) and turbulent dissipation (ε). The turbulence is assumed to be of a pure diffusive nature, so that the unknown Reynolds stress tensor in the Reynolds-averaged Navier-Stokes equation can be described by a turbulent viscosity term. The turbulent viscosity υ_t is defined by k and ε:

$$\upsilon_t = \frac{c_\mu k^2}{\varepsilon}. \tag{A.1}$$

By using a Schmidt number of 1 it is equal to the diffusivity D_t. The equations for the turbulent dissipation and turbulent kinetic energy yield:

$$\frac{D\varepsilon}{Dt} = \frac{\partial}{\partial x_j}\left(\frac{\upsilon_t}{\sigma_\varepsilon}\frac{\partial \varepsilon}{\partial x_j}\right) + \frac{\varepsilon}{k}\left(c_1 P + c_2 \varepsilon\right), \tag{A.2}$$

$$\frac{Dk}{Dt} = P + \frac{\partial}{\partial x_j}\left(\frac{\upsilon_t}{\sigma_k}\frac{\partial k}{\partial x_j}\right) - \varepsilon, \tag{A.3}$$

where P represents the production of kinetic energy, given by:

$$P = -\overline{u_i' u_j'}\frac{\partial \overline{u_i}}{\partial x_j}. \tag{A.4}$$

The constants are determined from experiments in decaying homogeneous turbulence and a turbulent wall flow, which results in (Launder and Spalding, 1974):

$$c_\mu = 0.09,$$

$$\sigma_k = 1,$$

$$\sigma_\varepsilon = 1.3,$$

$$c_2 = 1.92,$$

$$c_1 = 1.44.$$

At the solid walls, it is assumed that the flow is parallel to the wall and that the computational domain starts at a small distance δ_w from the real wall, so that the computational domain starts in the logarithmic layer. At the inflow boundary, values for k and ε are prescribed. At the outflow boundary, a convective flux for k and ε is prescribed equal to zero (Neumann boundary condition).

LARGE-EDDY SIMULATION

The LES resolves the large-scale turbulent eddies (length scale on the order of the mesh size and larger), whereas the small-scale turbulent eddies are modelled using a subgrid-scale model. A filtered Navier-Stokes equation is therefore solved with an additional subgrid-scale stress term, which accounts for the unresolved turbulent length scales by means of a subgrid-scale eddy viscosity. The commonly used Smagorinsky subgrid-scale model was implemented here, for which the subgrid-scale viscosity is given by:

$$\upsilon_{\mathrm{SGS}} = C_s^2 \Delta^2 \left|\widetilde{S_{ij}}\right| . \tag{A.5}$$

where C_s represents the Smagorinsky constant, Δ is the filter length based on mesh properties ($\Delta = (\Delta x \Delta y \Delta z)^{1/3}$), and S_{ij} represents the rate of strain tensor based on the filtered velocities. At the solid walls a Van Driest damping function was used to let the subgrid-scale viscosity go to zero in accordance with the turbulent fluctuations. The Smagorinsky constant was taken as $C_s = 0.1$. Complex geometries can be implemented by an immersed boundary method (IBM), which forces velocities to be zero at the outer wall of the system. More details on the LES model and the applied IBM method are found in van Balen et al. (2009).

Appendix B

UV irradiation model

CALCULATION METHOD

For the calculation of the UV irradiance by the MSSS model (Liu et al., 2004), the UV lamp is divided into J different segments. At every node of the computational mesh, the path of the optical pathways from this node to each segment is calculated (Figure B.1). The angles θ_1, θ_2 and θ_3 are defined by Snel's law. The refraction angle θ_1 is calculated from:

$$s_1 + s_2 + s_3 = r_1 \tan\theta_1 + r_2 \frac{n_a \sin\theta_1}{n_q\sqrt{1 - \frac{n_a^2}{n_q^2}\sin^2\theta_1}} + r_3 \frac{n_a \sin\theta_1}{n_w\sqrt{1 - \frac{n_a^2}{n_w^2}\sin^2\theta_1}}, \tag{B.1}$$

where r_1, r_2 and r_3 are the radial lengths (m) of the air layer, quartz layer and water layer (distance from quartz-water interface to the computational node), respectively, which are all known. The summation $s_1 + s_2 + s_3$ is known and represents the horizontal distance (m) between the lamp segment and computational node. For the refractive indices of air, quartz and water, commonly used values equal to $n_a = 1$, $n_q = 1.54$ and $n_w = 1.33$, respectively, were used (Liu et al., 2004). The angle of refraction θ_1 is solved from eq. B.1 by iteration. The irradiance I_j from segment j to the computational node is calculated from:

$$I_j = (1 - R_1)(1 - R_2)\frac{P/J}{4\pi\,(l_1 + l_2 + l_3)^2} T_w^{\frac{l_3}{0.01}} T_q^{\frac{l_2}{0.01}} A_f \cos\theta_1, \tag{B.2}$$

where l_1, l_2 and l_3 represents the path lengths (m) in the air, quartz and water layer, respectively and J represents the total number of segments. The term $\cos\theta_1$ yields a correction caused by the cylindrical segments (Bolton, 2000). R_1 and R_2 are the reflectances at the air-quartz interface and quartz-water interface, respectively. Due to refraction, a focus effect occurs, because the radiant power is concentrated to a smaller cross-section than in the absence of refraction, which is represented by the focus factor A_f. According to Liu et al. (2004), the focus factor is obtained by dividing the cross-section area of a circular segment without refraction by that with refraction, resulting in:

$$A_f = \frac{(l_1 + l_2 + l_3)^2\,\Delta\theta\cos\theta_1}{r\,(h_2 - h_1)\cos\theta_3}, \tag{B.3}$$

where $\Delta\theta$ represents a finite difference angle, r is the sum of the radial lengths r_1, r_2 and r_3, and $h_2 - h_1$ represents the horizontal distance between two optical pathways starting from an angle of $\theta_1 - \Delta\theta$ and $\theta_1 + \Delta\theta$, respectively. Summation of the irradiance intensities for all the segments j yields the total irradiance in the computational node.

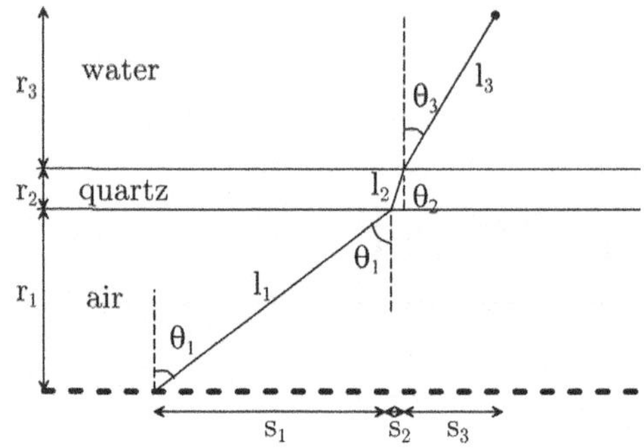

Figure B.1 Sketch of the refraction of an optical pathway emitted by the UV lamp

INTEGRATION OF UV IRRADIANCE OVER PARTICLE PATH

The fastest way to calculate the UV dose for each particle is to interpolate the irradiation field for each particle position. The irradiation field is obtained by calculating the irradiation at each mesh point on the finite-element mesh, as described above. Here, the irradiation (MSSS model) only needs to be calculated once for each mesh point, typically of the order of 100 000 times.

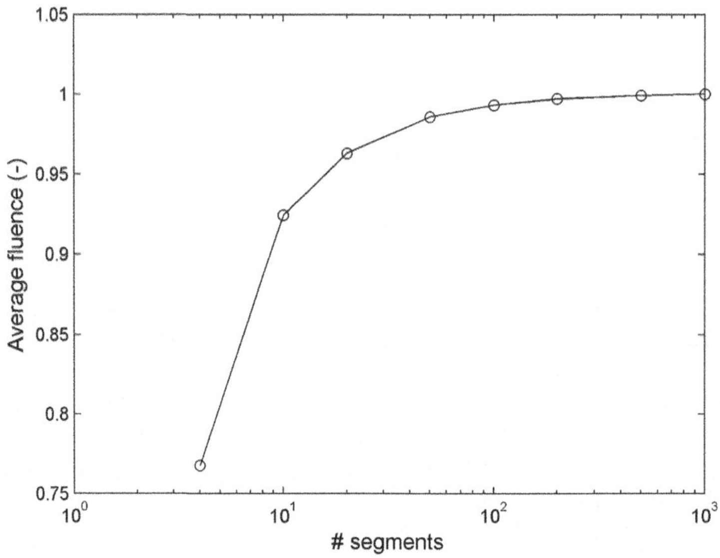

Figure B.2 Convergence for the number of segments used in the irradiance model.

A more precise method is to calculate directly the irradiation at each particle position. For all the particles, at every time step the irradiation needs to be calculated, which is of the order of 10 million (10 000 particles times 1000 time steps), a hundred times more expensive as the previous method. Both methods gave the same dose distribution for a single cross-flow UV lamp system, so the first method was used.

CONVERGENCE NUMBER OF SEGMENTS

The fluence distribution around a single UV lamp (radius 25 mm, quartz layer of 1.9 mm, height of 150 mm) in a rectangular mesh (100 × 100 nodes, mesh size of 2 mm) was evaluated for different numbers of segments. In Figure B.2, it was shown that an accurate solution for the average fluence is obtained (within 1% of the more accurate solution) after 100 segments.

List of publications

JOURNAL PUBLICATIONS (peer reviewed)

Wols, B.A., Hofman, J.A.M.H, Uijttewaal, W.S.J., Rietveld, L.C., Stelling, G.S. and van Dijk, J.C. (2008). Residence time distributions in ozone contactors. *Ozone: Science & engineering*, 30(1):49–57. http://dx.doi.org/10.1080/01919510701759538

Wols, B.A., Hofman, J.A.M.H, Uijttewaal, W.S.J., Rietveld, L.C. and van Dijk, J.C. (2010). Evaluation of different disinfection calculation methods using CFD. *Environmental modelling and software*, 25(4):573–582. http://dx.doi.org/10.1016/j.envsoft.2009.09.007

Wols, B.A., Hofman, J.A.M.H., Beerendonk, E.F., Uijttewaal, W.S.J. and van Dijk, J.C. (2010). A systematic approach for the design of UV reactors using CFD. AIChE Journal 57(1): 193–207. http://dx.doi.org/10.1002/aic.12255.

Wols, B.A., Shao, L., Uijttewaal, W.S.J., Hofman, J.A.M.H., Rietveld, L.C. and van Dijk, J.C. Evaluation of experimental techniques to validate numerical computations of the hydraulics inside a UV bench-scale reactor. Chemical Engineering Science 65(15): 4491–4502. http://dx.doi.org/10.1016/j.ces.2010.04.013.

Wols, B.A., Uijttewaal, W.S.J., Hofman, J.A.M.H., Rietveld, L.C. and van Dijk, J.C. The weaknesses of a k-ε model compared to a LES model for the prediction of UV dose distributions and disinfection. Chemical Engineering Journal, 162(2). 528–536. http://dx.doi.org/10.1016/j.cej.2010.05.055.

CONFERENCE PROCEEDINGS

Wols, B.A., Hofman, J.A.H.M., Uijttewaal, W.S.J. and van Dijk, J.C. (2006). The effect of turbulent diffusion on the performance of ozone systems. In *Proceedings of the workshop developments in drinking water treatment modelling*, Delft.

Wols, B.A., Hofman, J.A.M.H, Uijttewaal, W.S.J., Rietveld, L.C., Stelling, G.S., and van Dijk, J.C. (2007). Residence time distributions in ozone contactors. In *World Congress on Ozone and Ultraviolet Technologies*, Los Angeles.

Hofman, J.A.M.H, Shao, L., Wols, B.A., Uijttewaal, W.S.J. and IJpelaar, G. (2007). Design of UV reactors by CFD: model development and experimental validation. In *Proceedings of the European COMSOL Conference*, Grenoble.

Hofman, J.A.M.H, Wind, D., Wols, B.A., Uijttewaal, W.S.J., van Dijk, J.C. and Stelling, G.S. (2007). The use of CFD modeling to determine the influence of residence time distribution on the disinfection of drinking water in ozone contactors. In *Proceedings of the European COMSOL Conference*, Grenoble.

Wols, B.A., Hofman, J.A.M.H, Uijttewaal, W.S.J., Rietveld, L.C., Stelling, G.S. and van Dijk, J.C. (2008). A particle tracking technique to estimate disinfection efficacy in drinking water treatment plants. In *6th International Conference on CFD in Oil & Gas, Metallurgical and Process Industries*, Trondheim.

Wols, B.A., Hofman, J.A.M.H., Rietveld, L.C., Uijttewaal, W.S.J., Stelling, G.S., and van Dijk, J.C. (2009). CFD Modelling of Ozone Contactors and UV Systems. In *High quality drinking water conference*, Delft.

Wols, B.A., Uijttewaal, W.S.J., Rietveld, L.C., Hofman, J.A.M.H, and van Dijk, J.C. (2009). Hydraulic optimization of a single UV lamp placed perpendicular to the flow direction by experimental and numerical techniques. In *IO3A – 19th World Congress & Exhibition*, Tokyo.

Wols, B.A., Hofman, J.A.M.H, Beerendonk, E.F., Uijttewaal, W.S.J. and van Dijk, J.C. (2009) Design of Hydraulically Optimized UV Systems using CFD. In *IUVA - 5th UV World Congress*, Amsterdam.

NATIONAL PUBLICATIONS

Wols, B.A., Hofman, J.A.M.H, Uijttewaal, W.S.J. and van Dijk, J.C. (2006). Numerieke modellering van desinfectieprocessen. *H_2O*, 39(12):36–39.

Hofman, J.A.M.H, Shao, L., Wols, B.A. and IJpelaar, G. (2008). Prestaties van UV-reactoren te voorspellen met modellen. *H_2O*, 41(19):98–100.

Wols, B.A., Hofman, J.A.M.H, Uijttewaal, W.S.J. and van Dijk, J.C. (2010). Computermodellering van stroming in drinkwaterinstallaties. *submitted to H_2O*

DATA PUBLICATIONS

Experimental and numerical data used in this thesis can be accessed online (through http://dx.doi.org).
All data:
http://dx.doi.org/10.4121/uuid:c1ac7344-1419-4398-ba13-c757551c303f
Movies of particle distributions in ozone systems (Figure 4.8):
http://dx.doi.org/10.4121/uuid:f284a6f2-04a2-431e-9896-4a04bd61ac6b
Movies of spreading of dye around a single cross-flow UV lamp (Figure 5.3 & 5.4):
http://dx.doi.org/10.4121/uuid:954f0e25-aa39-454d-a0fa-656f36bf72d1
Movies of advanced oxidation around a single cross-flow UV lamp (Figure 5.18):
http://dx.doi.org/10.4121/uuid:dfff9d45-75c4-455c-848a-d50bc4a09087
Movies of particle distributions in single cross-flow UV lamp systems (Figure 5.20):
http://dx.doi.org/10.4121/uuid:63ef98c8-5a01-442a-9caf-80106576a171
Movies of spreading of dye in a bench-scale UV reactor(Figure 6.4):
http://dx.doi.org/10.4121/uuid:06d6b6e3-3fe1-4502-825b-6627a5deaa5f
Movies of a LES model of a bench-scale UV reactor (Figure 7.1):
http://dx.doi.org/10.4121/uuid:474d9ab9-c354-4d18-87c9-69b02acedce3

List of symbols

Roman symbols

A	Absorbance	–
C	Concentration of chemical substance	$g \cdot L^{-1}$
CT^*	Normalised CT value ($CT^* = CT \cdot k_s / C_0$)	–
D	Diameter of cylinder or pipe	m
D	UV dose	$mJ \cdot cm^{-2}$, $J \cdot m^{-2}$
D_{scale}	Characteristic dose ($D_{\text{scale}} = \frac{2P}{Q\alpha} \exp(-1)$)	$mJ \cdot cm^{-2}$, $J \cdot m^{-2}$
D_{mean}, D_{min}, D_{10}	Specific UV doses	$mJ \cdot cm^{-2}$, $J \cdot m^{-2}$
D_t, $\mathbf{D}$	Eddy diffusivity	$m^2 \cdot s^{-1}$
D_L	Dispersion	$m^2 \cdot s^{-1}$
DE	Decimal elimination	–
E	Energy	J
E_{CFD}	Photon flux calculated by CFD	$mmol \cdot cm^{-2} s^{-1}$
F	Cumulative distribution function	–
I, I_0	UV irradiation, with (I) or without (I_0) absorption	$mW \cdot cm^{-2}$, $W \cdot m^{-2}$
L	Length	m
N	Concentration of micropollutants	$mol \cdot L^{-1}$
N_A	Avogadro's number	mol^{-1}
N_p	Number of particles	–
OX	Oxidation	%
P	Lamp power	W
Pe	Péclet number	–
Q	Flow rate	$m^3 \cdot s^{-1}$
R	Reflectance of unpolarised radiant energy	–
Re	Reynolds number	–
St	Strouhal number	–
T_w, T_q	Transmittance in water or quartz over 1 cm	–
T_{10}, T_{50}, T_{90}	10th, 50th, 90th percentile of residence time	s

T_h, T_m	Hydraulic ($T_h = V/Q$) or mean residence time	s
V	Volume	m^3
dW	Increment of white noise process	–
f	Frequency	Hz
f	Distribution function	
h	Water depth	m
k	Turbulent kinetic energy	$m^2 \cdot s^{-2}$
k_s	Ozone decay rate	s^{-1}
k_μ	Inactivation rate	
k_i	Reaction rate of contaminant i with $OH\bullet$ radicals	$L \cdot mol^{-1} s^{-1}$
l	Optical path length	m
n	Refraction index	–
p	Probability density function	
$r_\perp, r_\parallel$	Reflectance in perpendicular and parallel direction	–
r_{lamp}	Radius of lamp	m
t	Time	s
u, v, w	Velocity in x, y, z-direction	$m \cdot s^{-1}$
$\overline{u'u'}, \overline{u'w'}, \overline{w'w'}$	Reynolds stresses	$m^2 \cdot s^{-2}$
u_m	Mean velocity	$m \cdot s^{-1}$
u_*	Shear velocity	$m \cdot s^{-1}$
x, y, z, r	Co-ordinate	m
$[N_i]$	Concentration of chemical substance	$mol \cdot L^{-1}$

Greek symbols

α	Specific absorbance ($\alpha = \frac{-\ln(T_w)}{0.01m}$)	m^{-1}
ε	Turbulent dissipation	$m^2 \cdot s^{-3}$
ε	Molar extinction	$L \cdot mol^{-1} cm^{-1}$
θ	Angle of optical pathway	–
θ	Normalised residence time ($\theta = t/T_m$)	–
κ	Von Karman coefficient ($\kappa = 0.41$)	–
λ	Wave length	m
λ	Life expectancy	s
λ	Distribution factor exponential pdf	–
ν	Viscosity	$m^2 \cdot s^{-1}$
ν_t	Eddy viscosity	$m^2 \cdot s^{-1}$
Φ	Quantum yield	–
ϕ	CT value or UV dose	

Abbreviations

ATZ	Atrazine
CFD	Computational fluid dynamics
CFL	Courant-Friedrichs-Lewy number
CT	Concentration-time (dose)
DNS	Direct numerical simulation
DMS	Dyed microspheres
FI	Fluorescence intensity
LA	Lagrangian Actinometry
LDA	Laser Doppler anemometry
LES	Large-eddy simulation
MMA	Micro-mixing analysis
PHBA	4-hydroxybenzoic acid
RANS	Reynolds-averaged Navier-Stokes
RTD	Residence time distribution
SFA	Segregated flow analysis
TKE	Turbulent kinetic energy
UV	Ultraviolet

IWA Publishing's authorised EU representative for General Product Safety Regulations is Diane D'Arras, 15 rue Duret, 75116 Paris, France, e-mail: safety@iwap.co.uk.

Printed and bound by CPI Group (UK) Ltd, Croydon, CR0 4YY
12/05/2026
02108891-0015